Nature™ Inc.

CRITICAL GREEN ENGAGEMENTS

Investigating the Green Economy and Its Alternatives

Series Editors
James Igoe, Molly Doane, Dan Brockington, Tracey Heatherington,
Bram Büscher, and Melissa Checker

Nature™ Inc.

Environmental Conservation in
the Neoliberal Age

Edited by Bram Büscher, Wolfram Dressler,
and Robert Fletcher

THE UNIVERSITY OF
ARIZONA PRESS
TUCSON

The University of Arizona Press
© 2014 The Arizona Board of Regents
All rights reserved

www.uapress.arizona.edu

Library of Congress Cataloging-in-Publication Data
Nature inc. : environmental conservation in the neoliberal age / edited by Bram Büscher, Wolfram Dressler, and Robert Fletcher.
    pages cm. — (Critical green engagements: investigating the green economy and its alternatives)
    Includes bibliographical references and index.
    ISBN 978-0-8165-3095-3 (hardback)
    1. Nature—Effect of human beings on. 2. Human ecology. 3. Conservation of natural resources. 4. Neoliberalism. 5. Environmental protection. I. Büscher, Bram, 1977–
II. Dressler, Wolfram Heinz. III. Fletcher, Robert, 1973–
    GF75.N363 2014
    363.7—dc23

                                                2013040334

Manufactured in the United States of America on acid-free, archival-quality paper containing a minimum of 30 percent postconsumer waste and processed chlorine free.

19   18   17   16   15   14   6   5   4   3   2   1

Contents

Part III. Nature on the Move: The Global Circulation of Natural Capital

Nature™ Inc.

Nature™ Inc.

The New Frontiers of Environmental Conservation

Robert Fletcher, Wolfram Dressler, and Bram Büscher

The global conservation movement is undergoing profound changes. While the venerable fortress conservation paradigm has been thoroughly critiqued (Brockington 2002; Igoe 2004; Adams 2004), the community-based conservation (CBC) approach that aimed to replace it, along with the integrated conservation and development projects (ICDPs) in which CBC is commonly based, has suffered a similar fate over the last two decades (e.g., Wells and McShane 2004; Dressler et al. 2010). Indeed, the very compatibility of conservation and development has recently been called into question by conservationists contending that the trade-offs between livelihood and environmental concerns may be largely irreconcilable (e.g., McShane et al. 2011). In response, the conservation debate has seen a myriad of divergent calls for alternative strategies, such as total landscape approaches (e.g., Sayer 2009) and a (partial) return to strict protectionism (e.g., Oates 1999; Terborgh 1999).

But while many novel hybridized versions of the older conservation paradigms are emerging in practice, the field of conservation also seems to be "reinventing" itself in its entirety to a degree that is not yet clearly understood. What is clear, however, is that this reinvention is very much tied to and in line with broader dynamics in neoliberal capitalism (Igoe, Neves, and Brockington 2010). This convergence is represented by mechanisms such as ecotourism, payments for ecosystem services, and biodiversity derivatives, as well as a variety of novel financial and technological instruments such as species and wetlands banking, carbon trade, and conservation social media, among others. With wildlife populations and biodiversity riches threatened the world over, new and innovative methods of addressing these threats are necessary—and none, we are told, are newer and more innovative than those drawing and/or relying on "the market." As public funding for conservation grows scarcer and organizations increasingly turn to the private sector to make up the shortfall, market forces have found their way into conservation policy and practice to a degree unimaginable only a decade ago. With much at stake, it is critical to investigate how such "neoliberal conservation" (Igoe and Brockington 2007; Büscher et al. 2012) is reshaping human-nature relations fashioned over two centuries of capitalist development.

Without going into depth here (but see the chapters by Dressler, Wilshusen, Igoe, and others for broader discussions), we see capitalist development as a powerful dynamic that originated in sixteenth- and seventeenth-century Europe and has since gone through a long and complex process of intensification, expansion, and struggle to encompass nearly all facets of life in virtually all areas of the world to a greater or lesser degree (Meiksins Wood 2002). This dynamic, in short, centers on a particular mode of production, circulation, application, and consumption that entails a continuous need for capital accumulation and growth of private profits. Neoliberalism, by contrast, has been enacted in earnest since the 1970s and refers to a particular ideology, governmentality, and set of practices that aim to replicate capitalist market dynamics across the social and public landscape (Fletcher 2010). Capitalism and neoliberalism, thus, are not the same and should not be confused conceptually (Foucault 2008; Fletcher 2010). Yet, they are intimately intertwined in that both thrive on and stimulate similar principles such as commodification, competition, financialization, and market discipline. While this introduction does not intend to give an in-depth analytical and conceptual exposé of the two (convoluted and complex) concepts, we stress some of their key elements as they relate to conservation. We start from the premise that the links between capitalism and conservation are long-standing (Grove 1995; Brockington, Duffy, and Igoe 2008) and that the links between neoliberalism and conservation also rest on more than thirty years of historical entanglement, conflict, and conjunction. Our aim in this introduction is to engage several key elements in these entanglements that have been central in the emerging debate concerning "neoliberal conservation" and that provide the overall structure for this volume supported by each chapter.

Three main lines of critical analysis, we contend, have dominated discussions of neoliberal conservation, and these will structure the different sections of this volume. The first line explores the ways in which neoliberal principles such as commodification, competition, financialization, and market discipline articulate with earlier conservation strategies, local socio-cultural dynamics, and rural livelihoods, producing novel mechanisms and major landscape changes in situ (Dressler and Roth 2010). The second line investigates the discourses, perceptions, and representations of neoliberal conservation and how they work to legitimate and "sell" novel relations between humans and nonhuman natures. The third line of analysis investigates the combined effects of these trends by assessing the mechanisms that transcend the conservation of particular in situ natural resources to allow for the abstraction and circulation of "natural capital" throughout the global economy. Considered together, these dynamics have produced a truly global conservation frontier: a suite of networks, activities, and regulations that are rapidly changing the relations between people and nature worldwide. This frontier traverses and connects the boardrooms of global hedge funds, trees owned by small farmers, consumers, interest groups, giant nature reserves, and a myriad of species (both human and nonhuman) trying to survive in changing ecologies. It boasts grand images of pristine landscapes connected to often contradictory material realities and consequences, leaving some actors struggling to access new markets and others dispossessed by various "green grabs" (see Fairhead, Leach, and Scoones 2012b). Building on Arsel and Büscher (2012), we refer to these new frontiers of neoliberal conservation as "Nature™ Inc."

Just as the frontiers of Nature™ Inc. are global and local, interconnected and dispersed, highly complex and ambiguous, so too have been academic efforts to understand them. Hence, despite the rapid proliferation and diversification of this literature, there remains a need to organize it within a cohesive analytical structure, to push critical analysis in new directions, and to map new arenas for future research beyond the bounds of current study. This is what the present volume offers, focusing on the three themes outlined above, shorthanded as (1) Nature™ Inc.–society entanglements; (2) representations of Nature™ Inc.; and (3) the global circulation of natural capital.

The remainder of this introduction lays the groundwork for this discussion, situating it within the history of research addressing the neoliberalization of environmental policy and practice, foreshadowing the interventions offered by our contributors. We begin by outlining the rapidly growing academic literature analyzing contemporary neoliberalism, describing how this analysis has been applied to environmental policy and in particular to describe the phenomenon we refer to as Nature™ Inc. We then trace the development of Nature™ Inc. over the past several decades, identifying a trend toward increasing abstraction and financialization in order to facilitate the global circulation of "natural capital" as the emphasis has shifted from ecotourism through payment for environmental (or ecosystem) services (PES) and on to newer mechanisms like carbon markets and species banking. We conclude by providing brief overviews of the chapters that follow, focusing on the three themes detailed above.

The (Mis)Uses of "Neoliberalism"

As Flew (2011, 44) observes, neoliberalism "has been one of the great academic growth concepts of recent years." From a mere handful of references in the 1980s, Boas and Gans-Moore (2009) identify a dramatic surge in scholarly attention to neoliberalism in subsequent decades: between 2002 and 2005 the term appeared in more than a thousand social science academic articles yearly. The concept's popularity has only increased in the intervening years, rendering its usage increasingly diffuse "such that its appearance in any given article offers little clue as to what it actually means" (Boas and Gans-Moore 2009, 139). At its worst, neoliberalism has become "nothing more than a vehicle for academics who like to criticise things that they do not like" (Igoe and Brockington 2007, 445), while at its broadest, the term is used "as a sloppy synonym for capitalism itself, or as a kind of shorthand for the world economy and its inequalities" (Ferguson 2010, 171).

This conceptual confusion is unfortunate, since neoliberalism, more precisely defined, reflects a distinct process with tremendous global influence, as many perceptive analysts highlight (see, amongst others, Peck 2010; Steger and Roy 2010; McCarthy 2012). As such, we and the contributors to this volume argue that it remains extremely important to (be able to) engage with the term. Growing complaints that neoliberalism has become so overused that it has lost all meaning—or even that the process it refers to does not actually exist as such—risk contributing to the hegemony of neoliberalism itself by allowing the ideology to fade from the realm of public discourse and insert itself as the invisible and hence unquestionable common sense of our time (Peck 2010; Büscher et al. 2012).

Neoliberalism defined more strictly is commonly identified with the widespread trend toward increasing relaxation of state oversight over political-economic affairs and reliance on the "invisible hand" of the market to efficiently allocate resources across the social landscape. Castree (2008a), building on Harvey (2005), characterizes neoliberalism as promoting the interrelated processes of decentralization, deregulation (or, rather, reregulation from state to nonstate actors), marketization, privatization, and commodification. This is not to imply a homogeneous and static process. Rather, it implies that neoliberalism diffuses sporadically, unevenly, in articulation with local sociocultural patterns and institutions (Harvey 2005; Foucault 2008; Dressler and Roth 2010; Steger and Roy 2010; Roth and Dressler 2012). Hence, many scholars have taken up the call to describe the diversity of "actual existing neoliberalisms" (Brenner and Theodore 2002; Duffy 2012) rather than positing some pure ideological edifice from which existing institutions are presumed to deviate (see esp. Brenner and Theodore 2002; Roth and Dressler 2012). Likewise, as a partial, uneven, and ongoing process, analysts increasingly speak in terms of neoliberalization rather than neoliberalism per se (see esp. Peck 2010). Yet, as Brenner, Peck, and Theodore (2010, 332) point out, "empirical evidence underscoring the stalled, incomplete, discontinuous or differentiated character of projects to impose market rule, or their co-existence alongside potentially antagonistic projects (for instance, social democracy) does not provide a sufficient basis for questioning their neoliberalized, neoliberalizing dimensions." Notwithstanding pronounced diversity in practice, similar dynamics can be observed in a wide variety of contexts, informed by a coherent set of theoretical prescriptions (Harvey 2005; Foucault 2008; Peck 2010) and manifesting in policies and practices with a distinct family resemblance. Indeed, neoliberalism's very flexibility can be seen as one of its most essential characteristics (Peck 2010; Duffy and Moore 2010).

So characterized, neoliberalism/neoliberalization has been analyzed in a number of interrelated ways. Most broadly, there is a strong distinction between treatments inspired by neo-Marxist and poststructuralist thought, respectively (Castree 2008a; Ferguson 2010; Fletcher 2010; McCarthy 2012; Wacquant 2012; Overbeek and Van Apeldoorn 2012). It is primarily in terms of their analysis of the nature and motives of neoliberal governance that the two perspectives diverge. In Harvey's (2005) paradigmatic Marxist reading, neoliberal *economics* is an ideological smokescreen concealing a more fundamental class project of accumulation by dispossession aimed to employ free-market policies for private appropriation of the commons. Hence, Harvey asserts, "It has been part of the genius of neoliberal theory to provide a benevolent mask full of wonderful-sounding words like freedom, liberty, choice, and rights, to hide the grim realities of the restoration or reconstitution of naked class power, locally as well as transnationally, but most particularly in the main financial centres of global capitalism" (119). Similarly, Overbeek and Van Apeldoorn (2012, 4) define neoliberalism as a political project aimed to restore capitalist class power in the aftermath of the economic and social crises of the 1970s and the challenge posed to the rule of capital globally by the call for a New International Economic Order.

For Foucault (2008), by contrast, neoliberalism is a broader approach to human governance in general, a particular "art of government" or "governmentality." More than a class project or ideology, then, in Foucault's reading, neoliberalism—particularly in the US context—is a "whole way of thinking and being," a "general style of thought, analysis and imagination" (218). In contrast to conventional understandings of governmentality (see Rose, O'Malley, and Valverde 2006), describing the processes by which subjects internalize social norms that compel them to self-regulate in the absence of overt domination, a specifically *neoliberal* governmentality operates through the construction and manipulation of the "external incentive structures within which individuals, understood as self-interested rational actors" (Fletcher 2010, 173), make decisions among alternative courses of action. Thus, Foucault (2008, 260, 271) describes neoliberalism as an "environmental type of intervention instead of the internal subjugation of individuals," a "governmentality which will act on the environment and systematically modify its variables."

The distinction between these perspectives is significant and should, pace Ferguson (2010, 171–72), always be spelled out. While the contributors to this volume adopt different positions in this debate, we take inspiration from several recent efforts to synthesize the two perspectives into a workable framework for analysis and potential action. Ferguson, for instance, suggests that "bringing these two different referents together can be more interesting, if we don't just equate them, but instead reflect on the conceptual themes they share (broadly, a technical reliance on market mechanisms coupled with an ideological valorization of private enterprise and a suspicion of the state), and use such a reflection to ask if the new arts of government developed within First World neoliberalism might take on new life in other contexts, in the process opening up new political possibilities" (173). Similarly, Wacquant (2012, 66, emphasis in original) proposes "a *via media* between these two approaches that construes neoliberalism as an *articulation of state, market and citizenship* that harnesses the first to impose the stamp of the second onto the third."[1] Springer (2012), among others, pursues reconciliation of the two perspectives as well.

In all, and despite their proliferation and sometimes dilution, debates and discussions concerning neoliberalism continue to display a richness and analytical sophistication that make wading through the conceptual muddle well worth the effort (see also, e.g., McCarthy 2012). Moreover, the contemporary situation makes it necessary to engage the term in one way or another in order to tackle the power dynamics that influence so many facets of modern life, including conservation. We describe next how these debates have played out concerning analysis of environmental policy, charting the emergence of debates concerning neoliberal conservation in particular.

From Neoliberal Environments to Nature™ Inc.

Following McAfee's (1999) prescient identification of the emerging trend involving "selling nature to save it," the critical academic literature addressing the process we refer to as Nature™ Inc. became centered on analysis of neoliberal *nature*. Important trailblazers were McCarthy and Prudham's (2004) seminal paper introducing a special issue of *Geoforum*, Heynen and Robbins's (2005) introductory article in *Capitalism Nature Socialism*, and the spate of writing, anthologized in *Neoliberal Environments* (Heynen et al. 2007), that followed. Alongside this work, the first efforts to conceptualize neoliberal *conservation* emerged (Sullivan 2006a; Büscher and Whande 2007; Igoe and Brockington 2007), which, analysts pointed out, demands unique mechanisms in order to harness the value of resources in situ (see esp. Büscher et al. 2012). While the neoliberal environments literature has continued to develop fruitful insights (see, e.g., Bakker 2009; Castree 2010a, 2010b), the neoliberal conservation discussion quickly expanded as well, soon producing several books (Brockington, Duffy, and Igoe 2008; Brockington 2009; Duffy 2010) as well as a variety of special journal issues (Brockington and Duffy 2010a; Sullivan 2010a; Arsel and Büscher 2012; Büscher and Arsel 2012; Roth and Dressler 2012; Fairhead, Leach, and Scoones 2012a; Corson, MacDonald, and Neimark 2013) and an array of individual articles too numerous to mention. Indeed, because of the importance, global presence, and urgency of the topic, this literature will continue to expand rapidly (assisted by the new book series on the topic initiated by the University of Arizona Press, of which this volume is the first installment).

While the present volume builds upon all of this previous work, it engages in particular with Arsel and Büscher's 2012 Forum issue of *Development and Change* in which the concept of Nature™ Inc. was first advanced. The authors highlight the three interrelated dimensions of this term ("nature," "trademarked," and "incorporated"), observing that it follows a long line of similar attempts to highlight the increasingly corporate nature of a variety of socioenvironmental processes designated by such monikers as Life Inc. (Rushkoff 2011), Green, Inc. (MacDonald 2008), Environment, Inc.

(Bosso 2005), and so forth. Yet they also highlight the double meaning of the "incorporated" qualifier to signify as well the fact that within neoliberal conservation "nature needs to be rendered a distinct 'corpus,' an 'entity' that stands outside of society and economy" (Arsel and Büscher 2012, 59). Meanwhile, the "trademarked" dimension of the term emphasizes the fact that within the framework of Nature™ Inc. the "nature" in question must be "protected, legalized, and institutionalized by particular systems of power and associated symbols" (Arsel and Büscher 2012, 60). Finally, use of the contentious term "nature" (see, e.g., Latour 2004; Goldman, Nadasdy, and Turner 2011) is intended not to designate some inert force external to human affairs but to highlight the intricate entanglement of humans and nonhumans within complex "socionatures" as well as to emphasize nonhumans' agency as "actants" in such networks rather than as the passive objects of human manipulation.

The *Development and Change* issue, in turn, was based on papers presented at an international conference held the previous summer at the Institute of Social Studies in The Hague, the Netherlands, from June 30 to July 2, 2011. This conference was the first large assembly of scholars seeking to critically interrogate "the market panacea in environmental policy and conservation," and it introduced the Nature™ Inc. concept to frame this trend. Originally intended to attract around 60 participants, the conference's call for papers drew over 230 submissions (of which 180 were accepted for presentation), further evidencing the groundswell of academic interest in this area.

This volume offers a further cross section of these conference presentations, carefully selected in order to provide a sample of the range of perspectives in the emerging literature and updated to reflect the most recent developments and offer predictions of future trends. The volume thus provides the first comprehensive critical overview of the full range of contemporary debates concerning neoliberal environmental conservation, drawing together the substance of many of the special journal issues and articles preceding it. Each of these addressed important yet specific aspects of the conversation, though all failed to provide a more structural overview of the debate. This volume goes further than merely providing a state-of-the-field review of the neoliberal conservation discussion, however; it also pushes the conversation in productive new directions by providing innovative theoretical work and empirical material in relation to what we identify as the three most significant new "frontiers" of environmental conservation: Nature™ Inc.–society entanglements; representations of Nature™ Inc.; and the global circulation of natural capital. Before introducing the various chapters and their specific interventions, we offer a brief overview of the development of Nature™ Inc. in order to historicize our discussion.

The Evolution of Nature™ Inc.

The commodification of natural resources is, of course, not a new phenomenon (Bellamy Foster 2000; Harvey 2006b; Nevins and Peluso 2008; Peluso 2012). The rendering of nonhumans as "fictitious commodities" (Polayni 1944) has occurred for at least as long as a capitalist mode of production has pursued its relentless quest to colonize new spaces, times, peoples, and processes across societies and landscapes (Harvey 1989, 2005). What is relatively recent, however, is the widespread effort on the part of capitalist industry to internalize natural resources as an integral component of production for "sustainable" management in the long term rather than simply externalizing environmental (as well as social) costs in the interest of short-term profit (Brockington, Duffy, and Igoe 2008). This is what Martin O'Connor (1994a) calls capitalism's "ecological phase," which can be seen to have commenced in earnest in the 1970s—the very period of neoliberal consolidation (Harvey 2005; Peck 2010)—with the acknowledgment of the environmental "limits to growth" (Meadows, Meadows, and Randers 1972) and the convening of the first major international conference (United Nations Conference on the Human Environment) in the same year as the publication of Meadows, Meadows, and Randers's text to confront this reality.

This coincided as well with a growing recognition of the human costs of traditional approaches to conservation, entailing state-centered "fortress" style management commonly prescribing the coercion and displacement of large numbers of resource-dependent peoples (or seemingly less confrontational resource substitutions) who were thus justifiably hostile to those responsible for their condition (see Wells and Brandon 1992; Peluso 1993; Neumann 1998; Brockington 2002; Igoe 2004; Dowie 2009). Out of this recognition grew the integrated conservation and development and community-based conservation campaigns. These campaigns sought to reconcile formerly competing concerns for conservation and development, incorporating the local peoples most dependent upon and knowledgeable about immediate resources as integral "stakeholders." Such interventions pursued a Hegelian synthesis of sorts between opposing theses, with development planners called upon to address environmental management and conservationists compelled to include human development in their work as well. MacDonald (2010b, 527) points toward the crucial role in this mission of the Convention on Biological Diversity, an "active political space . . . in which rights and interests may be negotiated and new social relations configured around those negotiations." He continues that "this arena can lead to creative opportunities for new, and previously excluded, groups to claim authority, but it also creates a context in which privileged positions and perspectives can be consolidated and codified in ways that structure policy and practice." This, he argues, is indeed what happened.

Central to this effort was the need to generate revenue from natural resources without substantially degrading them over time, and it was in this aim that the first seeds of Nature™ Inc. were sown. While resource commodification in the form of extraction and processing had been (and still is) seen as a relatively straightforward process, achieving the opposite—commodification through conservation, or what West (2006) calls "conservation-as-development"—required novel ways of thinking and performing. How could value be generated from resources preserved in situ when value had almost always previously been created by transporting resources from their place of origin and thus fleeing the localized environmental and social impacts effected by this displacement? A raft of novel institutional approaches was soon developed in pursuit of this agenda, with earlier devolved CBC strategies now taking on new market mechanisms to conserve "ecosystem services" by placing an imputed market value on them, the income from which would purportedly provide local users with incentives to curb the extensive use of natural resources (Dressler et al. 2010).

Of course, once again, this was not an entirely new phenomenon. As Brockington, Duffy, and Igoe (2008) contend, protected conservation areas, while commonly framed by proponents as bastions of pristine nature standing opposed to the base forces of predatory capitalism, have, in fact, always been connected with processes of capitalist commodification, particularly in the form of the nature-based tourism (e.g., safari trips, trophy hunting) commonly promoted inside them. Yet in the neoliberal age, this commodification has intensified and transformed to a degree unimaginable in those halcyon days of yore. Hence, one of the first moves of Nature™ Inc. was to magnify and transform this nature-based recreation—now relabeled "ecotourism"—as an ostensibly "nonconsumptive" (and thus sustainable) form of income generation. Other means of harnessing the value of in situ resources, from bioprospecting to ostensibly sustainable forms of resource extraction (i.e., logging), were promoted as well.

The chief problem with such mechanisms in terms of commodification is that the value they generate is fundamentally tied to the environments they address, requiring either the movement of people to the site of production (in the case of ecotourism) or the transport of resources to the site of consumption (bioprospecting, sustainable forestry). Moreover, scholars keep pointing out the enormous "gender costs" of these initiatives in terms of women's livelihoods and "lost spaces" (Harcourt 2012).[2] The "friction" (Tsing 2005) resulting from this movement increases transaction costs substantially, reducing both potential profit and the ecological gains such mechanisms ostensibly provide. Thus, a major innovation in the development of Nature™ Inc. came with the formulation in the 1990s of the payment for environmental services (PES) mechanism. This, of course, built on the growing framing of "nature" as a "service provider" in general, a perspective also promoted by the Convention on Biological Diversity initiated in Rio in 1992 (see Robertson 2006; Sullivan 2009; MacDonald 2010b) and popularized by ecological economist Robert Costanza and colleagues' (1997) ambitious effort to quantify all the environmental services provided by the planet. Through PES, "consumers" of ecological services could now pay their "producers" remotely, allowing for the spatial separation of consumers from the resources they (non)consume and thus a partial abstraction of value from any particular landscape, given that within the PES framework environments are rendered equivalent such that degradation in one location can be "offset" by preservation elsewhere (see Brockington, Duffy, and Igoe 2008; Sullivan 2009). This then initiated a bold new era in conservation, a shift from hybridized forms of CBC–Nature™ Inc. to increased reliance on so-called market mechanisms.

Yet even in PES there are important limitations to the accumulation process. Essentially, conservation is still tied to a particular piece of land, inhibiting the abstraction of value from dependence on any particularities of place and thus nonhuman natures' transformation into full-fledged commodities that could circulate globally (see esp. Büscher, this volume). Hence, Nature™ Inc. has truly come of age with the recent development of innovative financial mechanisms that facilitate this abstraction, separating the creation of value from connection to any particular environment and thus allowing value to circulate freely around the globe as fully fungible stores of value (see Bracking 2012; Sullivan 2013b; Büscher, this volume). The rise of the global carbon market, facilitated by the "flexible mechanisms" of the UN Framework Convention on Climate Change's (UNFCCC) Kyoto Protocol (also emanating from Rio), in which abstract carbon credits are traded between spatially disconnected players, is only one aspect of this trend (see Bumpus and Liverman 2008; Lohmann, this volume). Species and wetlands banking, in which destructive development in one area can be offset by purchasing credits ostensibly representing equivalent preservation elsewhere (see Pawliczek and Sullivan 2011; Sullivan 2013b), signifies its intensification. Environmental derivatives take Nature™ Inc. to new heights by trading not in any particular conservation mechanism at all, however abstract, but rather in markets only loosely linked by other conservation mechanisms to actual existing landscapes (Büscher 2010a; Cooper 2010). In this spirit, major international initiatives such as REDD+ have pumped millions of dollars into national treasuries with the hope of working through decentralized structures and local institutions to pay or provide (livelihood) cobenefits to users to avoid deforestation by sequestering carbon in anthropogenic forests. Facilitating and sustaining carbon sequestration via a range of market incentives for local behavioral changes now draws on, or works through, existing institutions from earlier interventions and dramatically reinforces the circulation of finance and capital that helped stoke climate change in the first place — effectively, new forms of carbon capitalism (Fletcher 2012b).

What is significant about this Brave New Neoliberal World is that, to an unprecedented degree, capitalism is endeavoring to accumulate not merely in spite of but rather *precisely through* the negation of its own negative impacts on both physical environments and the people who inhabit them, proposing itself as the solution to the very problems it creates. Büscher (2012, 29) thus characterizes neoliberal conservation as "the paradoxical idea that capitalist markets are the answer to their own ecological contradictions." This contradictory process signifies both a partial reversal of familiar capitalist engagements with nature (very partial, that is, because more "traditional" engagements continue and also seem to intensify) and an intensification of capitalist dynamics at the same time. It is this trend, both intensifying historical patterns and transforming them in important ways, that we call Nature™ Inc.

Through all of this, Nature™ Inc., like the neoliberalism that underpins it, continues its global ascendance. As MacDonald (2010a) relates, it was at the World Conservation Congress in Barcelona, sponsored by the International Union for Conservation of Nature, in September 2008 that serious debate concerning the appropriateness of market mechanisms and corporate partnership within the sphere of what Brockington and colleages (2008) call "mainstream conservation" was finally displaced to the margins. By the time of the Convention on Biological Diversity's 10th Conference of the Parties in Nagoya, Japan, in October 2010, such debate had all but disappeared as the campaign to calculate and create markets for trade in the Earth's environmental services culminated in the conference's uncritical endorsement of the United Nations Environment Program's (UNEP) The

Economics of Ecosystems and Biodiversity (TEEB) initiative (MacDonald and Corson 2012). As a "grounded" extension of this, these days it seems that the prime objective of conservation policy is to "grab green" locally, what Fairhead, Leach, and Scoones (2012b, 237) define as "the appropriation of land and resources for environmental ends." Hence, neoliberal conservation has become part of a discursive process manifesting materially as broader resource and landgrabs (White et al. 2012; Fairhead, Leach, and Scoones 2012b), the main goal of which is the appropriation of value of seemingly anything material and important as basic "inputs" for life. As such, there is less space for "nature" to function as its own actant: it is needed for its "services." All this is captured by the slogan "Nature is Dead! Long live Nature™ Inc." (Arsel and Büscher 2012, 53).

At the time of writing we are now precisely twenty years past Rio, and thus the future course of conservation is being charted yet again by the outcome of the Rio+20 conference held in June 2012 (as well as the subsequent World Conservation Congress in Jeju, South Korea, in September). Discussions at Rio+20 centered on the concept of the "green economy," advanced via a recent United Nations Environment Programme report (2011) as a purportedly novel replacement for a "sustainable development" increasingly pronounced dead on arrival. By all appearances, green economy discourse represents yet another intensification of Nature™ Inc. (Brockington 2012). At the same time, resistance to neoliberal conservation is growing, signified not only by the rise of the critical literature surveyed above endeavoring to problematize the trend but also by the increasing on-the-ground critique advanced by social movements throughout the world concerning the perils and pitfalls of increasing commodification of socionatures.

How this contest will play out remains to be seen. The chapters in this volume offer a variety of important tools to understand this situation, both now and as it develops and transforms into the future. In the next section, we explain how.

Outline of the Book

The book is divided into three sections. The first addresses what we call Nature™ Inc.–society entanglements. In the initial contribution to this section, Wolfram Dressler presents a fine-grained analysis of the ways in which conservation has become increasingly neoliberalized in Palawan Island, a biodoversity hotspot in the Philippines. Dressler grounds his analysis in the distinction between "first," "second," and "third" natures, where the first designates resources subject to only minimal manipulation and employed primarily for their use value; the second applies to landscapes that have been substantially transformed for commodity production (e.g., agriculture); and the third describes ways in which "commodity relations extend further to produce and assign an abstract market value to nature based in speculative assumptions, categories, and representations of how nature *should be* or *ought to become* for and by humans within and beyond local environments." Based on these distinctions, Dressler describes a movement from "first to third nature," contending that current conservation practices around Palawan's celebrated Puerto Princesa Subterranean River National Park are increasingly entangling both raw materials and commodity-producing landscapes within virtualistic speculative relationships that tie their value to global financial markets. The unfortunate consequence of this movement has been "to replace and erase a forest landscape filled with deeper history, cultural heritage, political relations, and social meanings."

In chapter 2, Ken MacDonald and Catherine Corson continue the section by analyzing the social construction of "natural capital" through the studies, programming, and practices of global biodiversity financing, focused on the emerging The Economics of Ecosystems and Biodiversity (TEEB) initiative. Drawing on Carrier and Miller's (1998) notion of "virtualism" and data gathered at quite a unique field site—namely, international conferences and meetings—they argue that, in line with neoliberal conservation, the consortium of actors orchestrating TEEB "abstract out" nature by assigning market values to it, "materializing" it, and rendering it "natural capital." This, they contend, unfolds within and is supported by international conservation institutions such as the Convention on Biological Diversity (CBD), which promotes abstract notions of nature as capital as the only basis for saving it. This promotion occurs not only through the market structures that assign financial value to material nature but also through "performance"—the process by which actors enact expertise and authority in the work of alignment and articulation of neoliberal abstractions of natures. In this sense, social gatherings like the CBD conferences entail more than just "exchange" of ideas; rather, they constitute an ideological space wherein the agency of TEEB is established and enacted as the basis for the production of natural capital as unquestioned "doxa."

In chapter 3, Frank Matose provides another one of the more empirically grounded analyses in the book. Looking at how local villagers around a protected forest in Zimbabwe negotiate and respond to various forms of dispossession, he draws attention to Gidden's idea of "structuration" in relation to neoliberal conservation. Structuration relates to the "reproduction of social systems," which in this case refers to a (post)colonial forest reserve that has long caused conflicts both among local communities and between these communities and the Zimbabwe Forestry Commission. Neoliberal conservation here is still very much "fortress conservation," with the Forestry Commission finding different ways to marginalize forest residents from the resources they previously utilized. Matose, however, goes beyond this top-down analysis to explicitly emphasize the agency of forest residents and the different types of resistance they employ to counter externally directed conservation practices. He distinguishes between overt and covert resistance in response to dynamics of privatization, the rollback of the state, and new forms of conservation-territorialization. He concludes that through these dynamics, different types of agency become visible in what are otherwise often generalized as processes of straightforward "accumulation by dispossession."

The book's second section pulls back from this ground-level focus to explore representations of Nature™ Inc. in the global arena. In chapter 4, Robert Fletcher takes a provocative turn by critically engaging what he describes as the common gap between the "vision" and the "execution" of neoliberal governance manifested in a widespread failure of neoliberal conservation strategies to "perform as intended." Drawing on diverse theory from Foucault to Žižek, he accounts for the perverse contradiction between the rhetoric and the reality in neoliberal conservation by suggesting that its broader "virtual vision" requires interventions that run counter to free-market activities in order to preserve, reinforce, and actualize the original goals of market-based conservation. Fletcher contends, further, that failure to acknowledge this "characteristic disjuncture between vision and execution in neoliberal conservation may take the form of 'fetishistic disavowal'" —a simultaneous admission and denial. The practical consequence of this analysis is to show that dominant beliefs in the value of capitalism are remarkably resilient in the face of the facts that contradict market outcomes; rather than amend their beliefs and behaviors to "fit the new facts," individuals have the uncanny ability to deny or explain away these facts so as to preserve the capitalist "truth regime" (in which capitalists' own wealth progresses as they disavow their policies' excesses and contradictions!).

In chapter 5 Dan Brockington focuses on a very different but equally important phenomenon in neoliberal conservation, namely, its increasing dependence on a particular element of the "persuasive forces of profit seeking": celebrities. As celebrities might not be the first thing that comes to mind when thinking about Nature™ Inc., Brockington highlights their importance by pointing out that despite the general shallowness associated with celebrities, they also exercise substantial power and thus need to be taken seriously. Moreover, celebrities are part of a "broader evolution of the role of style in society in producing and reproducing distinction, privilege, and inequality." Conservation and environmental action are often also seen as mechanisms of distinction—for commercial, ego, or other purposes— while they help celebrities "look good" by "doing good." Indeed, Brockington shows how celebrities and NGOs are increasingly "doing good" together and so reinforce each other's image and commercial potential. He goes further to frame this dynamic in terms of Colin Crouch's idea of "post-democracies," or "disengagement and apathy with respect to politics by much of the citizenry that democracies are meant to empower. They are characterized by increasing inequality but apparent popular acquiescence to this fact." Celebrity endorsement, then, becomes a seemingly nonpolitical way of dealing with politically contentious issues of environment and biodiversity conservation. In the process, Brockington argues, conservation becomes spectacularized and part of the circulating capital on which the new "green economy" depends. He concludes that we have good reasons to be wary of these trends but equally stresses that "we need to be alert to the radical potential of celebrity." Celebrities are not uniform and might thus perhaps be able to join radical causes.

In chapter 6, Peter Wilshusen argues that we need to look at the "ways in which the term 'capital' has been discursively extended and transformed as a means of articulating concepts and organizing practices related to sustainable livelihoods and institutional design/environmental governance." Further, he contends that the "discursive extension of the term 'capital' beyond its original meaning within economics" in fact "erases" power by masking "an incremental process of economic reductionism within conservation/ development theory and practice." In building this argument, Wilshusen adds to the neoliberal conservation literature by rereading, rediscovering, and criticizing work by Bourdieu. On the one hand, he argues that we need to use Bourdieu's concepts of misrecognition and dissimulation to understand the way discourses about economic concepts are twisted to suit capitalistic ends. On the other hand, he points out that, ironically, this very understanding by Bourdieu was subsequently taken up by economists to do exactly that: to misrecognize the implications of key concepts and consequently assimilate them to neoliberal logic. In this case, the famous "five capitals" (physical, cultural, financial, human, natural) often used in the livelihoods literature illustrate the point. Wilshusen concludes by pointing at the broader applicability of the arguments made: misrecognition and dissimulation happen at both local and global levels and are similarly visible in current attempts to construct a "green economy."

In this section's concluding chapter, Larry Lohmann offers an incisive deconstruction of financial markets that capitalize, commodify, and trade the "natural" elements of disasters as fungible commodities. He describes the origin and logic of international carbon markets as the dominant market-based "solution" for reducing emissions for climate change mitigation, highlighting how science-based policy has been drawn into this mechanism, which supposedly enables the "internalization" of global warming "externalities" through the accumulation and trading of carbon under a capitalist profit motive. The limits of carbon trading to offset and reduce overall carbon emissions are effectively exposed by the *deep fallacy and contradictions* of the carbon market itself: how it makes concessions to the petroleum industry; how it is powerless to incentivize industry away from fossil fuel use; how it draws on private-sector responsibility to sustain and support the "buyer's market" for carbon (facilitating its exchange beyond the state by leveraging its price per ton); and how this last strategy will simply not work, as private-sector investment calls for clearly defined exchange markets that yield sustainable profits (with a clearly defined commodity, which carbon simply is not). In short, Lohmann argues that the processes by which carbon markets are said to internalize the negative externalities of climate change effectively extend captalist tendencies to extract, commodify, and accumulate capital assets for enhancing profit and reinforce a technocratic, neoliberal hegemony that only exacerbates climate change impacts.

The third and final section of the book explores the different ways in which Nature™ Inc. circulates throughout the global sphere, a dynamic we call Nature on the Move. In the first installment, Bram Büscher theorizes the process by which conserved nature is commodified within capitalist markets such that value can be abstracted and detached from its connection with particular in situ environments and thus "freely" exchanged throughout the global economy. Büscher observes that while previous research has effectively described the ways in which nonhuman nature is impacted by commodification, this research has largely neglected the converse dynamic in which nature is taken up and utilized by capital to become what Marx called "value in process." This takes the form of what Büscher labels "fictitious conservation," in that the value generated becomes increasingly difficult to connect with the actual resources ostensibly being conserved and from which this value supposedly derives. In the process, Büscher concludes, dynamics of production, consumption, and circulation become increasingly intensified, accelerated, and entangled in ways that make the conventional distinctions among them difficult to sustain.

In chapter 9 Jim Igoe builds on Büscher's analysis to elaborate on the fictional aspect of this "fictitious conservation." Drawing on Dubord's discussion of "spectacle," Igoe observes a growing trend wherein the mainstream conservation movement promotes the circulation of spectacular imagery in support of what he calls "ecofunctional nature": environmental policy that claims to "optimize economic growth and ecosystem health" through market-based interventions. He describes how, through this process, a long-standing preoccupation with "nature for contemplation" in the form of protected areas reserved for enjoyment by societal elites has been transformed into a "nature for speculation" (a double entendre signifying the synchronization of financial markets and media representations) that incorporates elements of the previous approach while dramatically transforming the human-environment relationship at the heart of this imaginary. In closing, Igoe highlights what is obscured by this spectacle of ecofunctional nature, namely, the "micropolitics" of quotidian struggles for control over both the access to and the signification of the nonhuman natures at stake in all of this.

This sets the stage for our final chapter, in which Sian Sullivan builds on both of the previous contributions to expand on this last theme. What Igoe's ecofunctional nature obscures most centrally, Sullivan contends, are all of the diverse ways of conceptualizing human-nonhuman entanglements that do not reduce either humans to rational actors calculating maximum utility or nonhumans to mere commodity exchange value. She points to a variety of other forms in which these relationships have been conceptualized both in non-Western societies and in premodern Europe, when animals were often accorded many of the same capacities as humans such that they could actively participate in legal proceedings (e.g., be prosecuted for murder). While acknowledging the potential pitfalls of excessive romanticization of indigenous environmental management practices (e.g., the common caricature of a particular people as "ecological noble savages"), Sullivan asserts that there are still valuable lessons that can be gleaned from such practices in pursuit of what she calls "(re)countenancing an animate nature" that is irreducible to an object of speculation in either financial or voyeuristic senses.

Conclusion

As is clear from the above, the book covers a wide range of topics and perspectives, providing the most comprehensive overview yet assembled of the rapidly growing study of neoliberal conservation. Despite its broad reach, however, there remain many issues that we have been unable to address here in as much depth as we would like. In particular, gender relations in neoliberal conservation are only treated tangentially and are indeed one of the glaring gaps in the overarching discussion as a whole. This, then, constitutes one of the important new directions to which this text points as a future research focus. There are several other potential topics we would like to note as well. One concerns the future of the green economy. At the time of writing, the concept has only just garnered serious discussion in the hallowed halls of global environmental governance, yet all signs suggest its growing hegemony in the years to come. How this latest gloss for Nature™ Inc. is performed, orchestrated, and circulated will thus be an intriguing research focus as well. As the increasing urgency of climate change pushes this issue ever further to the center of environmental governance discussions, this may have implications in terms of the issue's articulation and/or conflict with other issues (e.g., biodiversity preservation) that may provide useful research material as well. The enduring economic crisis is bound to have repercussions for the Nature™ Inc. agenda, yet these may go either way, either intensifying the agenda still further in the face of financial contraction or challenging it with evidence of its own limits. This prospect also warrants investigation.

These are (if we can be forgiven for using the term) mere speculations, however, and will doubtless be overshadowed by future events impossible to predict from our current vantage point. What is certain is that Nature™ Inc. in whatever guise is a force requiring urgent engagement by activists and academics alike. We hope that this volume helps to emphasize the importance of this agenda, to provide a state-of-the-art assessment of its current status as a scholarly field, and to offer productive and stimulating tools for understanding and researching the phenomenon both now and into the future.

Notes

1. But see critiques of Wacquant's position in Peck and Theodore (2012).

2. See the various chapters in Harcourt (2012), who argues in the introduction that the idea of livelihoods must be taken out of its dominant framing, as "gender aware, participatory, and just economic activities to sustain livelihoods are almost by necessity means of engaging in a critique of the prevailing economic system" (5).

PART I

Nature™ Inc.–
Society Entanglements

CHAPTER ONE

Capitalizing Conservation on Palawan Island, the Philippines

Wolfram Dressler

Market-based conservation is often heralded as the most efficient and effective means to improve rural livelihoods and conserve forests in much of Southeast Asia (Landell-Mills and Porras 2002; Pagiola, Bishop, and Landell-Mills 2002; Wunder 2008). State agencies, donors, and NGOs promote the payment of local users to maintain ecosystem services and further their involvement in market ventures to value alternate uses, reducing dependency on forest resources and extensive land uses (Pagiola, Bishop, and Landell-Mills 2002). However, as market-based conservation initiatives connect with local land-based production and consumption, various forms of resource partitioning, privatization, and commodification progress in the few remaining frontiers of the region (Rigg and Nattapoolwat 2001; Cramb et al. 2009; Potter 2009; Akram-Lodhi and Kay 2010). This chapter examines how the broader and local processes driving rural transformations have been accelerated through the progressive revaluing of nature in terms of market-based conservation on the frontier island of Palawan, an international conservation and development "hot spot." I demonstrate how the pace and scale of agrarian change between rural peoples in central Palawan has increased with the onset of resource partitioning, objectification, commodification, and revaluing through translocal "capitalist conservation." Changing management structures at the island's flagship protected area, the Puerto Princesa Subterranean River National Park (PPSRNP), have intersected with the local political economy of agrarian change, accelerating new capitalist investment, commodity production, and privatization amongst a mixed group of indigenous and migrant resource users. Drawing on the notions of first, second, and third natures, I show how market-based conservation and agrarian change converge to facilitate the parallel, overlapping shift from primary to tertiary production in one region. I describe how indigenous people in and using nature (first nature) for primary purposes such as subsistence become alienated from local resources as they work, shape, and extract from nature for commodity markets (second nature). I then focus on how such commodity relations generate an abstract market value on nature that yields equally abstract representations of idealized nature (third nature; Hughes 2005). I show that after decades of indigenous peoples using nature for commodity markets, the process of the national park being valued as a "common" World Heritage site has attracted new market investments that reinforce the coconstitution of conservation and agrarian change toward *intensified* capitalist production, rearticulating the value of nature on Palawan Island in an abstract sense according to modern images and ideals of the Philippines.[1]

From First to Third Nature: Conceptualizing Commodity Landscapes and Capitalist Conservation

The notions of first, second, and third natures offer a suitably nuanced conceptual approach to examine how capitalist markets influence the valuing of nature through agrarian change and conservation over time and space on Palawan Island. Yet the shifting, recursive nature of such "changing natures" suggests anything but a linear trajectory moving from a static historical baseline of human-environment relations. First nature, or "society in nature," refers to rural peoples producing within nature and for themselves but often with modern means and methods that are pulled from second nature. While not a "fixed" historical moment isolated from trade networks, in first nature, social meaning and labor are part of use-value production, as commodities remain part of familial social relations and economy, though they are connected with financial markets (Smith 2008). Second nature refers to an environment "as worked by people and shaped by extraction, agriculture, markets and other anthropogenic factors" (Hughes 2005, 158). In second nature, unequal commodity relations can sharpen as rural peoples spend more time producing resources once reserved for family needs as money commodities for the markets of others, leading in time to social and economic differentiation, with poor peasant producers (drawing on first nature) providing surplus and labor for intensified commodity production

(Akram-Lodhi and Kay 2010). Third nature extends simultaneously from first and second natures. It involves the ways in which such commodity relations extend further to produce and assign an abstract market value to nature based in speculative assumptions, categories, and representations of how nature *should be* or *ought to become* for and by humans within and beyond local environments (Wark 1994; Hughes 2005). As the economies of rural agrarian societies mix with but are partly supplanted by tertiary production controlled from afar (e.g., ecotourism, businesses, etc.), we see the onset of third nature through the emergence of ideas and assumptions that value nature speculatively in abstract market terms. In this "conditional sense, ideas of third nature carry assumptions of predictions" of how humans should value nature in economic terms (Hughes 2005, 158).

Third nature also conveys a virtual character that is conditional (Hughes 2005, 158), a character that is conditioned by how our ideals and assumptions come to value nature in a speculative fashion through the images and ideas of nature circulating in "markets" that inculcate us. Such abstracted nature is discursively produced, disembedded, contained, and subject to economic valuation. Rural areas zoned as protected areas and transformed into ecotourism hot spots, for example, draw people, beliefs, and representations that project both social and economic values and meanings onto a "virtual nature" that conforms to "models of economic abstraction, which are taken to be the fundamental reality that underlies and shapes [their] world" (Carrier and Miller 1998, 1–2). Eventually, investors, tourists, and locals who adopt and invest in such abstractions of nature are "said to perceive a virtual reality, seemingly real but dependent upon the conceptual apparatus and outlet that generate it" (Carrier and Miller 1998, 2). The dominance of such models suggests that all things in an "abstracted nature" are being commodified and perceived according to synthetic representations of nature and society that adhere to certain norms, beliefs, and meaning (Carrier and West 2009). Investing in such virtual nature amounts to a "new," less grounded reality wherein knowing, understanding, and acting on idealized representations of nature are manifest in needs, wants, and desires that, as bounded and labeled, can be consumed, reproduced, and projected as simulations (images, ideas, etc.) that come to represent reality without ever having to be in "nature" (Carrier and Miller 1998).

As third nature is produced and consumed by locating new values and meanings on how first nature should be (Wark 1994, 199), its valuing and consumption can have sustained social, economic, and biophysical implications for rural landscapes and peoples (West and Carrier 2004). Third nature becomes partly prescriptive when local reality is expected to conform to the meanings, signs, and symbols of abstracted models and when people act in ways that make this reality conform to the structures of abstract reality (Carrier and West 2009, 7). What evolves then are "collections of images, ideas, discourses and values that reproduce the material world according to ways that they imagine it to be" (Brockington, Duffy, and Igoe 2008, 193), giving rise to a commodified, stylized landscape that conforms to people's needs, values, and expectations of nature—particularly in terms of Western representations (West and Carrier 2004, 485).

The process by which first, second, and third natures unfold through time has become locally manifest on the island of Palawan, the Philippines. The sections below describe how translocal processes that include (1) agrarian differentiation, (2) the rise of commodity landscapes, and (3) shifts from subsistence to service-sector production have supported and reinforced third nature through capitalist conservation at *sitio* (hamlet) Sabang, Barangay Cabayugan, near the Puerto Princesa Subterranean River National Park.

Agrarian Differentiation, Commodity Landscapes, and Capitalist Conservation on Palawan Island

While the Philippines experienced a steady pace of agrarian change, central parts of Palawan only began "filling in" after World War Two. A steady influx of migrants from resource-scarce and/or violence-prone areas in the Philippines traveled to Palawan to find relative tranquility, abundant fishing grounds, and fertile lands already occupied and used by indigenous peoples such as the Tagbanua, Batak, and Pala'wan. In time, migrants from Luzon, the Visayas, and Mindanao comprised much of the local population, comingling and competing with indigenous peoples and producing settlement patterns consisting of intensive agriculture and commercial fishing in the coastal zone abutting the central cordillera. With just 35,369 people in 1903, the island's population grew to 755,412 people in 2000, with a population growth rate of 3.6 percent (Eder 1999; Philippine Census 2000). As the population grew, so too did infrastructure, concessions for timber and nontimber forest products, the release of alienable and disposable lands for homesteading, land titling through cadastral zones, and commercial agriculture near Puerto Princesa City, the provincial capital (Eder and Fernandez 1996; Eder 1999). Such agrarian change spread inward and upward.

The mid-1960s until 1986 were witness to Ferdinand Marcos's political influence, policies, and "investments" controlling forests, land, and people in central Palawan. The island's upland rain forests were exploited extensively through "logging, mining, [and] corporate farming rights, including rattan and almaciga . . . concessions" that overlapped with Cabayugan (Ocampo 1996, 37). Deforestation rates continued apace, and migrants from Bulinao and Cebu had settled in Cabayugan when in 1971 Marcos officially declared the island's 3,901-hectare flagship protected area, the Puerto Princesa Subterranean River National Park, in order to retain timber and tourism potential for his treasury.[2] In 1986 the first People Power Revolution ousted Marcos, restoring liberal democracy in the Philippines (Vitug 2000). Indigenous rights and environmental NGOs soon shifted their focus to Palawan, calling attention to indigenous land rights and deforestation on the island (Vitug 1993). For the first time, the convergence of agrarian change, land rights initiatives, and market-based conservation emerged through similar political-economic structures at the national park, reinforcing the recursive trend of first nature moving through to second and third natures along the prevailing capitalist currents of the time.

First to Third Natures: The Rise of Capitalist Conservation at the Puerto Princesa Subterranean River National Park

First Nature: Tagbanua Settlers, Commons, and Landholdings

The Tagbanua Apurhano settled Cabayugan well before migrant farmers arrived, park boundaries were zoned, and market-based mechanisms facilitated third nature through capitalist conservation. Settling in the 1800s, most held the well-forested area as common property. Originally from Aborlan and Napsaan, the Tagbanua of Cabayugan are characterized as a near-coastal people who had relied on marine resources, forest products, and swidden for subsistence, culture, and worldview for centuries (Fox 1954; Warner 1979). In the area, most Tagbanua long-traded forest products extensively with merchants and the neighboring Batak, "hunter-gatherer" people (Venturello 1907). As families moved inland, they made swiddens and moved on to fertile valley lands and coastal inlets farther northwest in Cabayugan, including areas near St. Paul Bay, land uses that defined Cabayugan's first nature.

Those Tagbanua settling near St. Paul Bay tended to define and regulate access to the commons according to religious sanctions and group membership. The presence of a *panya'en* and other malicious spirits in dream worlds mediated initial forest clearings; and access to landholdings and forest resources was partly restricted to those belonging to the Tagbanua community, loosely defined by ethnicity and blood ties. Preexisting political structures and leadership further influenced local decision making concerning which resources could be accessed and used, decisions based less in coercion than social suasion. Tagbanua cleared swidden, planted tree crops, cut nontimber forest products for sale, and grazed carabao on these lands with relatively few restrictions (see Warner 1979, 35).

In the early 1900s several Tagbanua pioneers were busy extracting usufruct plots from the commons. They planted coconut trees (*nyog*) and other permanent crops such as cashew (*kasoy*) and coffee (*kape*) for future subsistence and commercial sales on the foreshore area, Sabang Beach. The early Tagbanua pioneer Juan Francisco, for example, planted dozens of juvenile coconut trees flanking Sabang Beach, while his relatives planted coconut palms in Martape just across the valley. Closer to an area called Malipien, the old-timer and ladies' man Gorgonio Pangican was also busy clearing forest for swidden, again, planting tree crops in his field.

The labor these pioneer Tagbanua expended to clear lands was the basis for securing customary rights to their land—a fact recognized by both indigenes and migrants in what was then, seemingly, "society in nature"—first nature. Access to forest resources and swidden fields in Cabayugan was unrestricted until an expenditure of labor resulted in clearing and planting permanent crops and/or marking trees, which, advertising occupancy and use rights, secured a farmer's claim to the harvest and land (Warren 1977, 56). As these families cleared land, plots were segregated from the forest commons, becoming the founder's property. As such, Francisco's coconut trees at St. Paul Bay advertised his family's "use right" to an improved area—landholdings that would be inherited through successive generations. A farmer's right to the land and its products was thus bound to the labour he or she invested during field preparation (McDermott 2000).[3]

In the process of settling near Sabang, senior Tagbanua shaman (*babalyan*) with specific religious functions also assigned cultural beliefs to unique landscape features in the area. At the mouth of the underground river, Tagbanua harvested tuna (*thunnus*) in abundance, calling the area Tuturingen. Rather than enter the underground river, they paid homage to a *panya'en*, *ungao* (a malicious anthropophagus deity) occupying the recesses of the cavernous (and now world-famous) underground river system. While Tagbanua long engaged in the trade of nontimber forest products (Venturello 1907), families were generally in control of exchange, effort expended for sales, and effort in subsistence production with fellow indigenes in the area; the essence of first nature or "society in nature"

(Smith 2008). The Tagbanua of Cabayugan already held clear property rights on the flat, fertile lands at *sitio* Sabang and assigned cultural beliefs to a variegated karst landscape, the central feature around which migrants would settle and the national park would be delineated.

Second Nature: Arrival of Migrants, Commodification, and Differentiation

The arrival of migrant settlers in the 1950s until today opened up and transformed the Tagbanua commons into a commodified landscape, or second nature, supporting commercial agriculture, private title, and, ultimately, major tourism investments around the national park and providing the raw material, labor, and infrastructure for park management and tourism development years later. The initial pioneers landed in the 1950s and 1960s, with others arriving later in Cabayugan's central valley due to resource scarcity, unequal tenant-landlord relations, and conflicts in their place of origin (see Kerkvliet 1974; Chaiken 1994). Most started off with livelihoods similar to those of Tagbanua.

Other migrants soon followed upon hearing of the island's bounty from relatives who had already settled; some received assistance from family, often settling on lands cleared by their predecessors. Next-of-kin and prominent migrant pioneers also provided recently settled migrants with different livestock and farm implements, which facilitated settlement—a kin-based advantage for many expert paddy farmers. Still others with different backgrounds and skills (teachers, carpenters, etc.) settled using their own finances. The original migrant population was relatively homogeneous and hunted, fished, collected, and farmed alongside Tagbanua (Fox 1954; Conelly 1985). The area was land-rich and people-poor.

While the social relations between and within migrants and indigenous peoples were never entirely egalitarian, relatively common reciprocal work relations were soon succeeded by commodity relations spurred on by agricultural intensification, landgrabbing, and, eventually, the onset of land privatization—the commodity basis of second nature. Socioeconomic differences soon sharpened within and between each group. Locally powerful migrants set out to grab lands that indigenous farmers had already claimed by clearing forests for swidden and planting tree crops and, when lands were flat, for paddy rice cultivation. Wealthier migrants also hired poorer Tagbanua (and poorer migrants) to help clear forests for swidden and paddy farming on the flat, fertile alluvial lands in the central valley. As Tagbanua labored and produced goods for migrants, production and exchange relations supported fledgling commodity markets, unequal trade relations, and the release of land for private titling, with the Bureau of Lands releasing 204 and 500 hectares of land in Sabang and Nasuduan, respectively (see Dressler 2006).[4] Collectively, these lands were classified as free patent and were subject to intermittent cadastral surveys for full privatization in each *sitio*. The tree crops, trees, streams, and hills Tagbanua used as boundary markers were rendered obsolete with cadastral zones and, soon, all-encompassing national park zoning. Cabayugan's second nature had arrived, imbricating first nature.

Third Nature: A National Park and an Emerging Commodity Landscape

The declaration of PPSRNP at 3,901 hectares in 1971 and as a World Heritage site in 1999 intersected with and supported migrant settlement patterns, tourism increases, and infrastructure development, exacerbating differentiation, commodification, and privatization of land and nature more broadly, except now according to abstract market values based in speculative assumptions, categories, and representations—third nature.

In the 1970s and 1980s, prominent migrants who secured posts as managers and rangers used their political and economic leverage to employ relatives at the park and claim lands illegally in the coastal *sitio*, Sabang, near St. Paul Bay and Malipien, adjacent to or inside of national park boundaries (see Dressler 2006). In the 1970s, in particular, migrants living in Sabang whose wealth had grown due to paddy farming soon realized that the lands near the beach abutting the national park were worth a lot more money, unleashing a beach-front land acquisition frenzy. The land acquisitions began when a migrant named Locila Avecino purchased, for a meager P1,500, 1.5 hectares of Tagbanua Juan Francisco's lands and coconuts abutting the beach and national park. In turn, Rudolpho Tagburos bought 5 hectares of land from the first Tagbanua, Maruang (after baptism, Valentine de los Santos), who held beach-front lands under a registered tax declaration certification—de facto evidence of land occupancy and use. The going price was P3,000 (US$68) per hectare. Rather than stop there, however, farmers such as Avecino and Tagburos kept on buying Tagbanua land, with each farmer consolidating smaller plots of land on either side of the main market road for about 800 meters down from the beach. Each farmer secured about 11 and 24 hectares, respectively, for paddy rice cultivation, which they then subdivided amongst their children—private holdings that would support tourism and Cabayugan's emerging third nature.

The type of economic venture on these lands changed with a rise in population, land scarcity, park popularity, market-based interventions, and the growth of tourism ventures, beliefs, and attitudes. In the 1980s prominent members of the migrant community in Sabang and *sitios* farther south had negotiated park management initiatives to acquire the best lands around the park and support their own paddy rice agriculture. In the 1990s, with the park's devolution from the Department of Environment and Natural Resources (DENR) to the city of Puerto Princesa, local migrants loyal to the mayor, Ed Hagedorn, were given employment privileges at the national park. Many migrant park employees and others benefiting from park activities used their income to diversify their livelihood activities, investing in but increasingly shifting from paddy farming to start restaurants and hostels that catered to the growing number of domestic and international tourists. Moreover, to ensure his political mandate of a "clean and green" Puerto Princesa city, the mayor and his council enacted policies to prohibit the burning and clearing of old-growth forest for swidden in order to maintain a "nature aesthetic" of lush, intact forest canopy along the main road tourists (and locals) traveled to reach the national park. By now a combination of lush forest, paddy rice, and tourism infrastructure spanned and flanked this market road, spreading laterally along the beach front in Sabang, most of which was on Avecino's and Tagburos's landholdings—lands that Tagbanua once held in common.

NGOs soon capitalized on constitutional provisions made in 1987 for indigenous land claims (Certificates of Ancestral Domain Claim [CADCs]), granting Tagbanua, Batak, and mixed-marriage couples de facto tenurial security through these CADCs in 1996 and 1997 in *sitio* Kayasan and Cabayugan, respectively. With mixed groups of migrants occupying low-lying foreshore and valley areas once occupied by indigenes, most Tagbanua had already moved hinterland from the park, market road, and tourism infrastructure.[5] In 1999 third nature was nigh, with the park's expansion to 22,202 hectares and complete rebranding as a World Heritage site under the title Puerto Princesa Subterranean River National Park, reflecting the mayor's control over the local landscapes. The expansion created new traditional and multiple-use buffer zones that engulfed the ancestral domains in which community conservation initiatives and migrant businesses flourished, respectively. With the park's global status growing, visitor numbers soared from 30,776 in 1998 to over 100,000 in 2009, and, in time, new forms of nature "branding" retained the green rhetoric appealing to tourists who imagined Palawan as the last pristine ecofrontier, the imagery of which now circulated globally via media and the Internet (Park Visitor Statistics 2009).

Privatization and the Shift to Tertiary Production

While most projects in the uplands of the traditional-use zone supported stabilizing swidden farming with tree crops, sedentary agriculture, and new livelihood pursuits (see Dressler 2009), the further intensification of agriculture and the growth of tourism in the multiple-use zone facilitated the shift to tertiary production. In multiple-use zones, park managers and planners supported state, NGO, and private-sector investments in projects that boosted (1) the productivity of paddy rice, (2) the securing of private title and small-scale businesses, and (3) the development of the tourism infrastructure.

Park-based livelihood projects supported migrant business opportunities and paddy rice by maintaining water flows, nutrient inputs, high-yielding seeds, farm implements, and market outlets for rice sales, reinforcing market-oriented, agricultural intensification. In particular, the UNDP-COMPACT programs (Community Management of Protected Areas for Conservation), implemented in 2004, geared NGO and state assistance toward supporting the intensification of paddy fields and/or wage labor (United Nations Development Programme 2004, 26). Moreover, many migrant households received support from the local farmers' cooperative and other park-supported people's organizations that enabled them to secure start-up loans for capital buildup (i.e., hand tractors, water pumps, etc.) to further mechanize production and invest in small-scale businesses. Many migrant farmers' relatively unfettered access to capital and infrastructure gave them a productive advantage to enhance farming and small-scale businesses that supported tourism development further down the road.

Many migrants used their recently secured private title as collateral to build capital investments, start their own businesses catering to local and tourist needs, and, in time, secure more land so as to expand their tourism-based enterprises in Sabang and surrounds. In Sabang, in particular, Tagbanua lands have recently become privatized under a titling scheme supported by the city government and national park, enabling migrant farmers to expand production, invest in small general stores, and develop larger businesses that cater to a growing tourism market. In 2009, for example, the Land Amortization Management Project in Sabang was in full gear, with predominantly, but not exclusively, migrant farmers and business owners providing evidence of boundary markers, tax declaration certificates, and other means of identifying and proving landownership so their holdings could be tendered as private title.

As the local and tourist population grew, Bulinao and Visayan farmers, owners, and tenants became increasingly focused on purchasing and privatizing land near the beach, closer to Sabang. The acquisition and use of such lands reflected two broader patterns: initial grabs and purchases in the 1960s for farming; and new arrivals purchasing plots from pioneer landowners, as well as larger landowners partitioning their plots into smaller parcels, which they then rented to poorer tenant migrants at higher prices.

Three migrant farmers who had arrived decades earlier followed a similar process of settling, claiming, and buying land from Tagbanua, clearing land for *kaingin* and then paddy rice, and, after saving enough money from rice sales, again purchasing or claiming land in Sabang in anticipation of tourist arrivals. One pioneer farmer, for example, sold all of his fishing boats because of his older age and now secures an income from *sari-sari* stores (a basic general store) and remittances from his children who work wage-labor jobs elsewhere. Similarly, another farmer who settled in a *sitio* just south of Sabang in 1989 began cultivating paddy rice on his own lot and in 1994 used his savings and networks to relocate to Sabang because of the good livelihood potential there. Upon relocating to Sabang, the family acquired a small plot of land (30 by 40 square meters) worth about P15,000 (US$344), which now holds a restaurant that offers Filipino food and a store where they "buy and sell" rice and other goods. While family members once sold local rice in Sabang, they now purchase most rice from markets or the government granary in Puerto Princesa City and then resell it at a higher price to locals and new resort owners. However, this family's relatives and children continue to harvest paddy rice in a nearby *sitio*. They then sell it at their restaurant/store in Sabang, showing how preexisting livelihood activities can support new, commercially oriented initiatives. Other farmers first worked as tenants to save up enough money to buy land and start tourism enterprises closer to coastal Sabang.

In other cases, Sabang's gradually expanding tourism industry has drawn newcomers to Sabang in search of livelihood opportunities linked to tertiary markets: wage labor at two new four-star resorts (see below), part- to full-time management of beach cottages (e.g., Mary's cottages), and a combination of fishing and/or farming opportunities to supplement their nascent businesses. This, in time, prompted larger landowners to sell their lands for megatourism development.

Compared to when land was usually purchased to support the expansion of paddy rice farming, new properties at Sabang were now bought and sold with the clear intent of owning and operating a business. The rise in the number of people who have migrated to Sabang to start up tourism-related businesses has also increased the scarcity and transaction costs of land in the area. Realizing that property prices have risen, many who claimed and secured significant hectarage as private title early on were now busy partitioning medium-size lots and renting out smaller parcels of land to poorer migrant tenants and, recently, selling several hectares of land (approximately 8–10 hectares) to two luxury resort developers, Sheridan Beach Resort and Daluyon Resort, for major sums of money. In particular, after Tagburos claimed and subdivided 24 hectares of land to his children, his sons and daughters took advantage of rising land prices to sell their portions (allegedly for P4 million, or US$86,000) to the Sterling Hotel Group, which in turn built a ninety-plus-room resort costing between US$100 and $711 a night. A slightly lower-cost land transaction unfolded with the Daluyon Resort—the start of Cabayugan, Sabang's third nature.

Major landowners also began partitioning their lots for tenancy arrangements. Different types of tenants with different types of businesses exist, but most rent their lands from the local landed elite who own the majority of land along the main market road from Sabang. In most cases, farmers who live farther down the road with secure tenure may have one family member renting from Tagburos smaller parcels of land that are closer to Sabang. Obtaining small- to medium-size loans from the Cooperative Land Bank (for which collateral is not needed), drawing on their savings from surplus rice sales, and/or receiving remittances from wealthier family members abroad, they build small stalls as "buy and sell" operations, cafés, restaurants, and souvenir shops. The closer they are to Sabang (and its beach), the more rent they pay Tagburos's family. Generally, those tenants with one or more family members generating a sustained cash income (e.g., as a park ranger or by working at the megaresorts) and who regularly farm paddy rice are most able to pay for higher-priced parcels of land close to Sabang and to greater numbers of tourists. The locally uneven political economy and management have kept Tagbanua peripheral to the capitalist currents emanating from and supported by livelihood projects, tourism, and the park's commercial profile. As such, those locals who already own the most land and hold advantageous positions in the local political economy (e.g., the national park) will realize the greatest benefits from the shift in primary to tertiary production—or, conceptually, the shift from first to third nature in Cabayugan.

Virtual Commodification

It was only after the mayor and his political campaign put the national park and the underground river on the map as an international tourism destination that the coastal landscape became commodified and revalued in terms of "hypermodernity," a new, imported reality. The social relations, meanings, and materiality constituting local first nature became more fully commodified, disembedded, partitioned, and valued in monetary terms. Concurrently, people project social and economic values and meanings onto nature that correspond to abstract, commodified representations of "nature and society" emerging from a modern sense of style, consumption, and material desire, concepts that often stem from Manila.

Since the park's declaration as a World Heritage site in 1999, the city's drive to promote the park globally, the growth in market-based conservation, and booming tourism numbers have all sped up the rebuilding and rebranding of the local area into a commodity landscape: what was once forestland is now private title hosting locally owned restaurants, and, more ominously, what was once the common coastal property of Tagbanua now hosts two new major tourism resorts. With land and capital now commodified, labor was the only item remaining. Tragically, it was young Visayan and Tagbanua children who toiled away daily manufacturing hollow blocks for one of the resort's foundation, among related tasks, for less than the daily minimum wage (P150, or US$3).

Buoyed by commercial agriculture and capitalist conservation, related tourism development, infrastructure, and consumerism have rendered the local people and landscape of Sabang as packaged consumer commodities. Tourists and locals identify with, value, and consume aspects of nature as consumer commodities that (are made to) conform to their expectations when, in fact, these commodities are part of a very different local reality. In Sabang, tourists and local people now view, appreciate, and consume popular consumer images and brands as if they were part of the preexisting landscape; for example, one popular fast food symbol, Jolly Bee, is shown floating into the mouth of the underground river. Marketers plant these images knowing that many domestic tourists from Manila and elsewhere value the sense of comfort they experience with the symbolic meanings that such brands convey—the Jolly Bee is happy, full of hamburger meat, and contentedly isolated from the harsher reality of forest living, something from which tourists are also isolated. Few realize that the bee is actually floating into a Tagbanua sacred site that is still inhabited by a malevolent *panya'en* who residents believe will inflict sickness on those who enter its sanctuary (see figure 1.1).

A few steps away, visitors see the newly erected Tribal Restaurant, with a statue of Mickey Mouse beckoning tourists to come eat "jungle fare," which is sourced from Puerto Princesa City and cooked to the modern Filipino palate (see figure 1.2). Again, both domestic and international tourists

Figure 1.1. The less than jolly side of ecotourism, Sabang, Cabayugan, Palawan (photo credit: Olivier Begin-Caouette).

are familiar with the friendly smile of Mickey Mouse and his beckoning hands—a representation of local nature that is at once modern Filipino and American Disneyfication. Closer to the coconut groves planted by Tagbanua pioneers are new reflexology and massage therapy stalls for the many tourists who arrive in vans from Puerto Princesa City for a daylong tour of the underground river. On the way up and down, tour guides occasionally discuss how the darker tribal Batak and Tagbanua still live in the forest without venturing out, generating an image for tourists of the area's indigenous peoples as primitive, at once part of but also separate from the park's prevailing "nature aesthetic"—a nature aesthetic maintained by controlling the clearing and burning of forest for swidden, deemed incompatible with idealized images of nature.

Tourists' perspectives also reflect (and invest in) third nature in a virtual sense. After visiting the area, tourists have the option of posting their perspectives on the park's virtual World Heritage site, with one visitor exclaiming, "We . . . enjoyed the untouched forest and beautiful flowers

Figure 1.2. A "tribal restaurant" with jungle fare in Sabang, Cabayugan, Palawan.

and the amazing construction of the stairs going up the mountain and the beautiful stones laid on the trails." Another states that "this cave of mystery resembles the mystery of life's journey as well. This cave is something that Filipinos should be proud of and share to the world."[6] Other tourists with whom I spoke in March 2010 stated similarly that Sabang and the national park were ideal places for "getting back to nature" and that, with unspoilt forest, the park truly reflected one of the New Seven Wonders of the World (see below). One couple just off the plane explained that while they did not want the area to be overdeveloped, they did hope that the infrastructure would eventually become more "modern," as it is on Boracay.

Yet tourists do not need even to physically enter the forest or the underground river in order to be in it. The park's landscape has become virtual, spread across placards, billboards, TV screens, in-flight magazines, and various major international websites (see figure 1.3). In the city itself, visitors see brightly colored images of the mayor and foreign dignitaries on massive placards superimposed over pictures of the underground river and the karst landscape "valuing" and "selling" the national park to new arrivals. Outside of Palawan, when visitors exit Terminal 2 of the Ninyo Aquino International Airport in Manila, they see large images of the underground river sparkling

Figure 1.3. Palawan, a virtually commodified tourism landscape.

with pastel colors plastered high on the wall next to the Marlboro Man riding his horse into the sunset. Moreover, in the ethereal world of the World Wide Web, virtual voters recently chose the underground river as a finalist to be designated as one of the New Seven Wonders of the World, showcasing its "natural beauty" and "God given splendor" virtually.[7]

The image of the "tribal other," the pristine and mysterious "mouth" of the underground river, and the surrounding wild jungle have become a virtual spectacle of nature through which capitalism commodifies the material and immaterial as glossy goods that tourists can readily consume (Brockington, Duffy, and Igoe 2008). The social and material construction of such abstract representations of nature, in which people enter a social mélange where reality and fantasy become indistinguishable, enables investment, consumption, and the fulfillment of desires without realizing that such constructions are partly disembedded from the reality that underpins them. Tourists have come to expect a clean, green, stylized, and romantic underground river and forest landscape. Sabang's first nature has become virtually translocal; it is abstracted, valued in market terms, and reified to conform to people's needs, desires, values, and expectations of nature. Yet the people who supply the places, cultures, and resources for this translocal hyperreality are incorporated and marginalized to the periphery of this emerging commodity landscape—the essence of third nature.

Discussion and Conclusion

This chapter has shown that much of Cabayugan's landscape has been irrevocably altered due to decades of incremental agrarian change, privatization, and differentiation through market-based conservation culminating in capitalist conservation. The rise of capitalist conservation has stoked and been supported by park management rezoning, "added value" livelihood support, the growth of small-scale businesses, the rise in tourism numbers, and the recent development of major resorts. In the process, we have seen privatization and differentiation displace and dispossess the local poor, control labor relations, and, ultimately, produce spaces of a highly abstracted nature. In these spaces, local labor value is sourced and material resources are destined to be produced for others who, concurrently, partition nature as objects and brands with economic value, conveying idealized representations and meanings of many things local.

In Cabayugan, we have thus seen an irregular transition from first to third nature. Smith (2008, 33) refers to first nature as including modes of capitalist production that are embedded in local resource production and still somewhat "internal to nature." While farmers and other resource users in Cabayugan have always negotiated and mediated nature with agency, in the past their social relations within nature were more "symmetrical" as they drew on forest products and land for familial subsistence production more than surplus production for external markets. In an idealized sense, social relations were based in use-value production above and beyond producing commodities for monetary exchange value (Smith 2008). While Tagbanua have long produced resources for sale in markets that were likely also structurally unequal, more of their time was devoted to producing subsistence for familial consumption and exchange with migrants, Tagbanua and Batak farmers (see Eder 1987). By extension, the onset of second nature involved local social relations increasingly being alienated from resource production and exchange—in essence, a shift of use value to exchange value for commodity production through agriculture and capitalist ventures supporting conservation (e.g., ecotourism such as boat trips to the underground river). Second nature, then, implies the rise of asymmetrical social relations "with nature" where the resource user begins to harvest resources as goods with a surplus value to be sold and exchanged for cash (White 1989). In cases where the Tagbanua became "laborers" to produce commodities for park-related tourism markets owned by wealthier migrants, many became socially alienated from the product of their labor, and increasingly so as more labor time, particularly that of child laborers, was invested in working for others' profit.

Increasing levels of competition and specialization in local livelihood processes arose for both agriculture and tourism development, driving competition for scarcer land, agricultural intensification, and commercial production. In time, the production of goods with surplus value, such as paddy rice, and other consumer products demanded more intense economic transactions between landlords and tenant famers that supported the shift toward tertiary production (e.g., ecotourism operators and restaurant owners), particularly through the development of major tourism resorts by foreign investors. As second nature sets in, then, we begin to see capitalist relations becoming increasingly manifest and pronounced in space, reconfiguring and hybridizing with (pre)existing social relations of production and exchange such that new ways of being, believing, and acting emerge in one area (Smith 2008).

The shift to a more abstract third nature thus involved the partitioning and economic valuation of nature in terms of ideas, images, and symbols that engendered and reproduced idealized representations of nature that came to simulate "real nature." We saw how in Sabang hyperreal notions of nature arose through the synthesis of multiple images and meanings in various settings—from airports to beach restaurants—that can structure tourists' and local people's sense of reality and truth about the "local world" despite often being divorced from actual reality. Tourists' longing for the national park's nature aesthetic helped to project both values and meanings onto nature that conform to "models of economic abstraction, which are taken to be the fundamental reality that underlies and shapes the world" (Carrier and Miller 1998, 1–2). Hughes (2005, 158) notes further that locally partitioned and economically valued nature can thus involve "speculation, rather than exploitation . . . [which] in this conditional sense . . . often carries assumptions or predictions regarding human action" in nature. Ultimately, then, Cabayugan's third nature entails both the tangibles and intangibles of nature being commodified within and beyond local landscapes, starting to replace and erase a forest landscape filled with deeper history, cultural heritage, political relations, and social meanings. While these place-based meanings have long mixed and merged with changing values and understandings, they tend to be forgotten when third nature consumes and removes much of first nature with relative force and degrees of permanence. While the notions of first, second, and third natures are conceptually abstract explanations of dynamic processes of agrarian change and capitalist conservation, the scale and pace of the transition are undeniable and will have profound implications for those people in nature apparently being "conserved."

Notes

1. I examined this trend by using a range of qualitative and quantitative methods in the area from 2001 until 2010. Data were collected through key informant interviews and participant observation of livelihood strategies, resource commodification, and the politics of capitalist investment in and through conservation. Household-level data on agrarian change came from a livelihood questionnaire in 2001 (N = 157) examining changes in livelihood, asset holdings, and land sales. Two follow-up questionnaires were conducted in 2006 and 2009 to identify livelihood changes among key indigenous swidden farmers (n = 20) and migrant business owners (n = 20), respectively, who were in the process of privatizing land or were share tenants on the lands of wealthier migrants in the tourism *sitio*, Sabang, near the underground river at the national park.

2. At the time, the national park was named St. Paul Subterranean River National Park.

3. Since land was clearly abundant prior to the 1960s, it was the availability of labor and not land that limited production (Warner 1979; McDermott 2000).

4. Key informant interview, Eduardo Castillo, Cabayugan Centro, spring 2004.

5. Much of the CADC lands were leased for twenty-five years as public domain and had no potential of being titled any time soon.

6. See http://www.worldheritagesite.org/sites/puertoprincesa.html.

7. See http://nature.new7wonders.com/28-finalists.

CHAPTER TWO

Orchestrating Nature

Ethnographies of Nature™ Inc.

Kenneth Iain MacDonald and Catherine Corson

In the summer of 2007 a reader of the London-based *Independent* posed a question to former US vice president Al Gore: "In 1992 you advocated a new set of 'rules of the road' for the conduct of the global economy, to take account of environmental costs and benefits. What progress do you think has been made since then?" Gore responded: "Not nearly enough. And actually, a re-examination of accounting systems and measurement protocols to include the environment in the routine, everyday calculations by which our economy is governed, comes about as close as you can get to the heart of why we have this crisis. . . . [A]ccounting systems are required to hold routinely in mind factors that are deemed to be important and significant in weighing the pros and cons of any decision. There has been progress to reform and redesign the accounting system. But not nearly enough."[1]

Gore's remarks were prescient. They were uttered just four months after a German proposal to undertake a study on the "economic significance of the global loss of biological diversity" had been adopted at the 2007 Potsdam G8+5 meeting. Three years later, in 2010, during a press conference that introduced the resulting study—The Economics of Ecosystems and Biodiversity (TEEB)—at the 10th Conference of the Parties (COP10) to the Convention on Biological Diversity (CBD), the team leader, Pavan Sukhdev, a former senior banker with Deutsche Bank and, until recently, head of the United Nations Environment Programme's Green Economy Initiative, made a striking comment: "This is one world; it's ours to create.

Let us create it and make it what we want, rather than wait for it to be dictated to us through further crisis and further problems."[2]

These comments from Gore and Sukhdev neatly reflect the rhetorical force of "natural capital." The world that TEEB seeks to create is one that materializes Gore's image of a nature simultaneously "accounted" for and made subject to market exchange. In many ways this attempt to bring nature into alignment with an expressed vision of that world is nothing new and reflects the process that Carrier and Miller (1998), among others, have described as "virtualism."

"Nature" has always been brought into being through processes of abstraction—ways of cognitively imagining, or more often being taught to imagine, one's surroundings as existing in particular ways for particular reasons such that they can be acted upon toward particular ends. Through time and across space people have imagined "nature" in different ways, with accordant differences in what were considered legitimate modes of interacting with the world around them. However, these ways of imaging the world have rarely been uniform or gone uncontested. Even in instances where ideological domination assumed doxic (or taken-for-granted) qualities, there have been competing modes of abstraction. Consequently, the conditions that created the dominant abstraction, and the practices of enacting it, needed to be continually (re)enforced.

In the past two decades, a particular image of "the world" as natural capital has gained prominence. In some sense this is not new. In industrialized societies "nature" has been implicitly treated as capital. What is new is a striking reduction in the opposition to the idea of a natural world defined as capital. Environmental institutions such as the CBD that might have challenged the subordination of "nature" to "the economy" have rapidly become strong proponents of market-based mechanisms through which nature is being increasingly privatized, commercialized, commodified, commoditized, and ultimately enclosed and, in the process, erasing pre-existing socionatures (see Brockington and Duffy 2010b; Büscher 2009; Carrier and West 2009; Castree 2008a, 2008b; Heynen et al. 2007; Igoe, Neves, and Brockington 2010; McAfee 1999). These processes not only have given rise to the concept of "ecosystem goods and services" but are also actually creating markets for their exchange (Robertson 2007; Sullivan 2011a), a process essential to their materialization.

What interests us here are questions about the dynamic processes whereby new markets and property relations are created and defined and in which power relations are realigned (McCarthy and Prudham 2004). How is natural capital enacted, and how are the conditions that create the abstractions upon which it depends produced and reenforced? This process, we suggest, requires the continual (re)alignment of actors, labor, and instruments around specific interests and ends. Further, that alignment involves substantive efforts of articulation (Hall 1986), circulation, and orchestration in attempts to enlist actors, institutions, and instruments in the project of (re)producing what we once knew as "the environment," or "nature," as "natural capital" (see Mitchell 2008).[3] While we see this as an integrated effect of neoliberalism, our focus is not on neoliberalism per se but on revealing the important role of performance and the enactment of expertise and authority in the work of alignment and articulation that neoliberalism (in all its variegated forms) requires.[4] We see that work as an important component of what Carrier and Miller (1998) describe as "virtualism."

In this chapter we combine the theoretical lens of virtualism with the empirical object of a new multilateral project (TEEB) and the physical site and instance of the COP10 to explore how processes of performance, orchestration, alignment, and articulation stitch together a dense weave of interests and actors in making real a vision of "nature" as capital. TEEB began as a study on the economics of biodiversity loss. While officially hosted by the United Nations Environment Programme (UNEP), TEEB's working units, including a communications hub and a scientific coordination group, were located in Germany and financed by the European Commission, Germany, the United Kingdom, Norway, the Netherlands, and Sweden. Led by Pavan Sukhdev, the project's goal was to produce a Stern Report for biodiversity.[5] As it unfolded, TEEB linked and mobilized a group of actors focused on the pricing and costing of ecosystems and biodiversity, producing reports aimed at distinct bodies of decision makers and putting in place demonstration projects oriented around mechanisms to incorporate the productive value of ecosystems and biodiversity into national accounts.

We argue that TEEB, which as a performative project mobilizes the alignments and articulations required to overcome obstacles to the realization of "natural capital," is an institutional expression of an environmental vision intended to bring the world into conformity with that vision (Carrier and West 2009). In what follows, we use our observations on TEEB to further refine the concept of virtualism, asserting that virtualism begins with an ideological commitment, in this case to place an economic value on nature. Yet, we also understand virtualism to be an ongoing process of reproduction grounded in conditions of contestation, where directionality emerging from the configuration of power relations and agencies is continually in the making. This means that any virtualism must be linked through virtual moments. It also demands that virtualism be performative: making the world conform to an image of itself requires constant orchestrating, aligning, and articulating actors, interests, institutions, and mechanisms to turn fragile social ties into durable associations (Latour 2005).

While the "performativity of economics" paradigm has been associated historically with studies of specific market technologies generated at specialized sites (e.g., Callon 1998a; Garcia-Parpet 2008; Holm 2008; MacKenzie 2003; MacKenzie, Muniesa, and Siu 2007), recent analysis of economic performativity explores the processes of "economization." This agenda is inclusive of a larger variety of sites and practices (Çalışkan and Callon 2009, 2010) than those generated at relatively local and specialized sites. With this in mind, we emphasize the performativity of a conference site, where the site itself serves as a stage that conditions the agency of TEEB in the production of natural capital as reality. In revealing the work of TEEB as performative in conforming reality to virtual reality by creating the conditions for the emergence of ecosystem markets, we highlight the importance of particular sites and spaces in the (re)production of agencements that we see as essential for an understanding of virtualism.

Virtualism: Conforming the World to an Abstraction

Carrier and Miller (1998) define virtualism as the attempt to make the world around us look like and conform to an abstract model of it. These abstractions, they claim, become virtualism when virtual reality stops simply being a description of reality and becomes prescriptive of what the world should be. The "set of partial analytical and theoretical arguments that define a world . . . becomes a virtualism when people forget that the virtual reality is a creature of the partial analytical and theoretical perspectives and arguments that generate it, and instead take it for the principles that underlie the world that exists and then try to make it conform to that virtual reality" (Carrier and West 2009, 7). Virtualism, then, "is a social process by which people who are guided by a vision of the world act to try to shape that world to bring it into conformity with their vision" (Carrier and West 2009, 7).

Miller (2005) discusses the correspondence between powerful actors, powerful discourses, and the degree of control they come to exercise over the world through their ability to be performative, and he distinguishes, for example, between more and less powerful actors, with the more powerful exercising "the ability to construct an economic world as the pure product of their own performativity, . . . reflecting their ability to take the virtual (i.e. the model) and actualize it in the world" (Miller 2005, 10). However, we argue that "realizing the vision" of natural capital does not involve distinctions between more or less powerful actors but rather requires bringing into being configurations of actors (in which we include devices, institutions, organizations), which become the reality they seek. It is the contestation among a multitude of actors, where power is relational, contingent, and dynamic, that is important.

In order to envision the ways in which actors and agencies are drawn together over time to enact the world, we draw on what Callon has termed an *agencement*—by which he means a heterogeneous ensemble of actors "made up of human bodies but also of prostheses, tools, equipment, technical devices, algorithms, etc." (Callon 2005, 4) and which he uses to "denote sociotechnical arrangements when they are considered from the point of view [of] their capacity to act and give meaning to action" (Callon and Çalışkan 2005, 24). Callon's perspective is helpful because it premises the effectiveness (i.e., its capacity to do work) of a proposition (e.g., natural capital) on the ability to draw together a corresponding sociotechnical apparatus.

Making the world conform to an image of itself is a long, messy, and conflicted affair involving the constant work of orchestrating, aligning, and articulating actors, interests, institutions, and mechanisms and the turning of fragile social ties into durable associations (Latour 2005). These processes—the construction of agencements—both require performance and are also performative (Hardie and MacKenzie 2007; MacKenzie, Muniesa, and Siu 2007). In essence, we see TEEB as an actor constituted by and constitutive of a dynamic agencement that works to (re)produce and reify nature as an array of goods and services subject to costing and that provides the institutional basis for creating and positioning markets as a privileged arbiter in the distribution of biodiversity and "ecosystem services." Virtualism, then, is a contested process that, like hegemony, is never complete, although it can be successful. For virtualism to be successful, we assert, "virtual moments" need to be linked together through the alignment of actors situated differently across time and space, where, in Miller's (2005, 10–11) words, "it is possible to write about the general history of virtualism and to carry out ethnography on the virtual moment." By paying attention to those virtual moments, we can observe practices of orchestration, alignment, and articulation in ways that integrate actors into a shared orientation within a developing and expanding network that subsequently works to create a world in accordance with models of how the world ought to be.

Studying the Field of Biodiversity Conservation

Indeed, a primary contribution of this work and that on which it builds is the extension of an attempt to ethnographically study an event like COP10 to understand these practices of orchestration, alignment, and articulation and how the site or the event works to constitute a virtual moment as one among many in a translocal field of organized conservation (Brosius and Campbell 2010; MacDonald 2010a). This focus on the event illuminates work that is often disaggregated in space and performed in bureaucratic sites resistant to direct observation (but see Corson 2010; Mosse 2006; Robertson 2010). It also allows us to examine the reconfiguration of power relations among key actors as well as the emergence, circulation, negotiation, and stabilization of idealized categories of biodiversity, which subsequently serve as vehicles for the realization of "natural capital."

Our capacity to do this, however, is grounded in a reconfiguration of methodological practice based on rethinking the notion of "the field" in conservation social science. This is required in part because of the intensive institutionalization of conservation practice and policy that has occurred in the past two decades through mechanisms of global environmental governance such as the CBD. The structure of the institutions of environmental governance that emerged out of the 1992 UN Conference on Environment and Development consolidated state authority, redirecting state and donor resources away from bilateral relations with conservation organizations and aligning them with the CDB program of work and the funding of that program through the Global Environment Facility (GEF), the financial mechanism of the convention. This consolidation of state authority under the guise of internationalism reconfigured power relations (MacDonald 2010b) and positioned the mechanisms of the convention, particularly its mandated meetings, as active political spaces—arenas in which interests could be negotiated, new social relations could be configured around those negotiations, and privileged positions and perspectives could be consolidated and codified in ways that structure policy and practice (see Strathern 2000).

We can think of these spaces, then, as what some management scholars refer to as field-configuring events (Lampel and Meyer 2008): affairs that temporarily bring actors together; construct arenas for demonstrating, displaying, and promoting perspectives, mechanisms, techniques, and practices; and provide the institutional context and opportunity to transform contestation into legitimate outcomes and shape disparate organizations and individuals into a "community" that shares a common meaning system

(Scott et al. 2000). Accordingly, we can see them as sites of "culture making," of sense making, and of learning how to make sense.

Attending to these events is important, as the emergence of transnational environmental governance, the consequent threat of regulation, and the accordant possibility of subordinating some interests in "the environment" have drawn previously separated actors together into spaces in which claims over "nature" and the ideological and material struggles that lie underneath those claims become not only unavoidable but more readily visible and subject to scrutiny (Latour 2004). Within (and beyond) these spaces, actors intentionally seek to give substance to the institutions and organizations engaged in environmental governance in ways that express that interest. These events, then, though not necessarily privileged, become important sites in which to compile accounts of these interests; they are places where the stakes of actors are articulated, where actions and associations formed in relation to those stakes become visible, where dissension within and between groups becomes apparent, and where contestation over the shaping of conservation policy and practice becomes clear. They provide an opportunity to observe encounters (e.g., huddles among delegates) and actions (e.g., gestures, tones of voice) that do not enter the official record. By being present at the site, we are able to record the process of knowledge being translated and to observe how it gains traction in relation to particular interests. We witness meaning as it is being made, challenged, transformed, and translated. And we are exposed to the agency of those involved in the process of structuring, orchestrating, and scripting the event.

Of course, this notion of tracking phenomena through time and space is not new (see Marcus 1995, 2000, among others), but it does call for imaginative modifications of methodological practice, particularly when those relations being tracked involve multiple actors appearing simultaneously in time-condensed spaces. The size of a Conference of the Parties makes it impossible for any one researcher to effectively cover the entire event or even track specific projects, like TEEB, as they are represented across the event. In response to these challenges, we have been involved with a group of scholars in the formation of an innovative approach to studying events that we term "collaborative event ethnography."

The goal of collaborative event ethnography is to adapt ethnographic practice to the spatial and temporal demands of the event. This means breaking from the conventional model of the lone ethnographer and the geographically contiguous community and working to realize the benefits gained from a group of observers jointly developing an approach to the study of the event, jointly developing the analytic frame for the research, training the collective team, working together around agreed-upon objectives, with shared guides to participant observation, common formats for recording observations, and modes of sharing the resulting field notes, recordings, transcripts, photographs, and images. In many ways this is designed to mimic the ways in which other groups such as conservation organizations seek to understand and influence the outcomes of field-configuring events.

In the case of COP10, the collaborative team involved seventeen researchers. Each member was part of smaller teams constructed around a matrix of themes and topics. The selection of themes (e.g., the tracking of market logics) reflected a combination of what we identify as dominant issues influencing current conservation discourses, based on our past research as individuals and our group experience at the World Conservation Congress. Topics were dictated by the COP10 agenda, which provided structure to the event. Small teams were made up of members aligning with at least one theme and one topic and were guided in their work by team leaders. A team of five researchers tracked the presence of market-based mechanisms and private-sector actors, but they aligned themselves simultaneously with a topic that allowed us to witness the distributed presence of a project like TEEB (where it was mentioned, how it was invoked, by whom, to what end, etc.), something that would have been impossible for a single researcher.[6]

As much as this chapter focuses on the presence of TEEB at COP10, it is important to note that what is presented here is also the result of our having tracked, through time and space, the way in which the promotion of newly "appropriate" modes of conceiving, making legible, and acting upon nature have gained credence and come to define a field or assume a strong "field mandate." Through the work of the team at COP10 it became clear that TEEB both symbolizes and enacts such a mode and is in the process of assuming a strong field mandate within organized biodiversity conservation.

The Virtual Moment of TEEB

Held once every two years, a Conference of the Parties is the primary meeting of the parties to the CBD. It is the venue where those parties revise text that was negotiated at the interim meetings of various working groups and advisory bodies to the CBD and render decisions based on that text. These decisions structure the program of work and the ideological orientation of the CBD. The conference draws together actors with an explicit interest in biodiversity conservation and configures power relations likely to mobilize material resources and institutional legitimacy in the continuing but shifting practice of biodiversity conservation. To be institutionalized within the CBD is to have the sanction of states and to be articulated with related institutions such as the GEF. The presence of authoritative actors with the capacity to implement mechanisms through their respective organizations and personal contacts helps to establish durable associations required for the realization of natural capital. Alignment and articulation as ongoing processes are key to the (re)production of those networks. It is this temporality that makes TEEB a moment in the virtualism of natural capital and the site of COP10 an instance in that moment, because the work of producing conformity must, almost by definition, enlist dominant institutional mechanisms and actors, which are revealed in the particular moments and at particular sites like COP10 (MacDonald 2010a).

The Alignment, Articulation, and Orchestration of TEEB: From Study to Approach

TEEB's capacity to generate alignment and articulation during COP10 was built on an existing institutional calculus put in place long before the meeting. The TEEB team had to bring together people with access to diverse sectors (e.g., politics, business, science, governance) and distinct sources of credibility. They had to design mechanisms for the circulation of information among individuals contributing resources to support the project, and they had to develop modes of communication that could both differentiate among these interests and maintain some degree of unified intent.

As TEEB grew from its origin as a proposal at the Potsdam G8 meeting in 2007 to an initiative, a study, and ultimately an approach, its structure took on new shape as various qualities and properties were used to align and articulate these different sets of actors. The best evidence of this lies in the composition of TEEB's fifteen-member advisory board, which includes key organizational leaders such as the executive director of the UNEP, Achim Steiner, and the director general of the International Union for Conservation of Nature (IUCN), Julia Marton Lefevre, together with leading ecological and environmental economists. Through the alignment of key academic and policy leaders, its embrace of so-called epistemic pluralism, and a diversity of economic instruments (Monfreda 2010), TEEB disembedded economic and policy expertise from their disciplinary and organizational confines and rearticulated them as allies in a common struggle.

In May 2008 the TEEB team released the first TEEB Interim Report at the ninth COP. By COP10 in 2010, the team had released five reports/

websites targeted to different audiences: ecologists and economists, businesses, national and international policymakers, local and regional policymakers, and citizens, whose website was titled Bank of Natural Capital (http://bankofnaturalcapital.com/). Unabashed about its intentions, a synthesis report (TEEB 2010, 4) states: "TEEB seeks to inform and trigger numerous initiatives and processes at national and international levels." It goes on to list various targeted processes and venues, including the G8+5 and the G20; the Millennium Development Goals; the 2012 Rio+20 Earth Summit; UN efforts to mainstream the environment in financial services; the Organization for Economic Cooperation and Development (OECD) responsible business conduct Guidelines for Multinational Enterprises; and industry voluntary guidelines.

The most striking evidence that TEEB is to become a key mechanism in state environmental planning and is likely to become an important "tool" in GEF's funding arrangements is found in several recommendations taken in intersessional meetings of the CBD in preparation for COP10. From the May 2010 fourteenth meeting of the Subsidiary Body on Scientific, Technical and Technological Advice (SBSTTA14), six key recommendations related to protected areas, sustainable use of biodiversity, and incentive measures explicitly advised parties and multilateral financial institutions, including the GEF, to look to TEEB for guidance in developing and implementing "additional means and methods of generating and allocating finance, inter alia on the basis of a stronger valuation of ecosystem services."[7] Two recommendations from the Ad Hoc Open-Ended Working Group on Review of Implementation of the Convention direct the executive secretary of the CBD to extend TEEB by working with UNEP, the World Bank, and the OECD to further develop "the economic aspects related to ecosystem services and biodiversity," develop "implementation tools for the integration of the economic case for biodiversity and ecosystem services," and facilitate "implementation and capacity-building for such tools." They also directed the secretariat to develop "capacity-building workshops, to support countries in making use of the findings of the TEEB study and in integrating the values of biodiversity into relevant national and local policies, programmes and planning processes."[8]

The Distributed Presence of TEEB at COP10

The use of COP10 as the stage for the rollout of TEEB attracted the attention and resource investments of potential affiliates. Within the confined space of a Congress Centre and over a concentrated time of ten days, TEEB's distribution system was able to reach the major influential actors across a range of ideological perspectives, encourage alignment, and publicize what actors no longer referred to as a study but as an approach. The discourse of natural capital was not, as in previous meetings, restricted to parochial discussions of economic incentives (MacDonald 2010a, 2010b). Instead, TEEB was well integrated across streams, making it difficult for any particular interest group to ignore. The heads of both UNEP and the CBD secretariat highlighted TEEB in the opening ceremonies, and it was a key presence in sessions devoted to ecological modelling, climate change, sustainable use, and parliamentary decision making, among others. In other words, it was widely distributed, widely promoted, and widely accessible.

That large plenary rooms — spaces in which large audiences could congregate — were reserved at particular times for TEEB-related presentations and that TEEB presentations were integrated into sessions organized by influential actors across the meeting indicated the intensity of the *work* that had gone into configuring a TEEB net*work* prior to COP10 with the specific intent of foregrounding it during the meeting. It relies upon associations with event planners or sponsors who have the capacity not only to "direct" through the configuration of spaces of presentation but also to integrate certain perspectives into a program in ways that achieve visibility and presence for that perspective. These associations, which were established well in advance of the meeting, enabled TEEB to have a distributed presence at COP10 so that the performance of TEEB could occur in front of a diversity of audiences.

This orchestration structures the performance of the model in ways that reveal power relations configured through the agencement. By observing and tracking the distributed presence of TEEB during COP10, we could observe TEEB as a political project — an agencement that extends beyond its intellectual substance. We could see virtualism unfold in practice where alignment and articulation drew actors together not simply by sheer force of material domination but through appeals to particular interests.

The Virtual Reality of Natural Capital

At the end of the formal presentation of TEEB to the parties to the CBD, as the applause was dying down and people were rising from their seats, a senior executive of a prominent UN agency leaned over to Pavan Sukhdev and, presumably not realizing that his microphone was still on, uttered the prophetic phrase "TEEB begins now!" This odd remark revealed COP10 as a critical turning point for TEEB. The transformation had multiple qualities: (1) TEEB was being institutionalized as a component of the CBD; (2) it was undergoing a metamorphosis from a study to an approach or a mechanism that would enable it to engage in performance and thereby draw more actors into its sociotechnical network; and (3) the results of its performance would create the conditions for the atomization and pricing of those "services" of nature not currently commoditized. To say that "TEEB begins now" suggests a shared understanding that what had occurred before the COP meeting was simply preparing the ground for the "real" work of TEEB.

As much as the performance of TEEB at COP10 can be analyzed as a virtual moment, TEEB did not begin at COP10 or at Potsdam. Indeed, TEEB and its role in the virtualism of "natural capital" *begin* with an ideological commitment to placing an economic value on nature—to remake nature into "natural capital," a vision that began long before COP10. Contrary to the currency that seems to be accorded to TEEB, it offers no new economic instruments: techniques such as green accounting and valuation and calls to internalize externalities, even as they were contested, have long defined the competing fields of environmental and ecological economics (e.g., Costanza and Daly 1992; Costanza et al. 1997; Daily 1997; Ehrlich and Ehrlich 1981; Pearce, Markandya, and Barbier 1989).

While there may be little new in the economics that TEEB invokes, what is new is the purposeful alignment of particular actors—an authoritative managerial class—brought together around it, the production of a visionary to serve as the embodiment of TEEB, and the calculated manner in which it has targeted key audiences. TEEB's institutional appeal lies in this sanction and the (re)packaging, (re)presentation, and (re)distribution of ecological and environmental economics as a product—an ostensibly implementable package designed, in relation to techniques of governance, to avoid complexity (and in doing so appeal to policy makers) and to easily adopt the reductionist managerial logics of "best practices" that accompany the operation and regulation of markets. If TEEB is the packaging and vehicle for performing economics, it is the longue durée of intellectual production that has produced a virtual reality of natural capital as an expressed image of the environment as a reservoir of capital, or "nature conceived in the image of capital" (O'Conner 1994a, 131). It is able to reproduce itself over time and space through the implementation of "regimes of investment" integrated in "a rational calculus of production and exchange" (Bellamy Foster 2002, 36). Accordingly, it presents environmental problems as failures to account for or adequately value (i.e., price and cost) components of nature. "The problem" in this vision is not with capitalism as a system of socioeconomic organization nor with markets as the basis for exchange and distribution but with a "nature" that has not been adequately priced.

In introducing TEEB at the CBD, Sukhdev described "the problem" as nature that has been economically "invisible." This invisibility is a shared problem with a shared solution: "The economic invisibility of nature must end. . . . Governments must respond to the economic value of nature by changing policies. . . . Companies must respond to the value of nature by recognizing their externalities and adopting a different and more responsible forward behaviour."[9] The solution, in accordance with this definition of "the problem," is to make nature visible as capital so that it can become part of the "rational system of commodity exchange" (Foster 2002, 35). Accordingly, realizing natural capital entails breaking the environment down into specific components—(ecosystem) goods and services—that can be alienated and brought into being as commodities, given an imputed price (TEEB would say a value), and subordinated to market mechanisms and policy instruments that use price as the basis for environmental protection.

TEEB and the Legitimation of Natural Capital

TEEB applies conventional practices of cost accounting to an "invisible" nature, simultaneously enabling other "market mechanisms" (e.g., PES, biodiversity offsets). As such, TEEB steps in to occupy sacred quantitative ground, providing the value determinations that "markets" and "payments" and "property swaps" require to be inserted into legal regimes of contractuality and moral spheres of equitable exchange. In this practice of accounting—or valuation—we enter the domain of Latour's metrology (1987, 15), in our case, the making of nature as a regime "inside which facts can survive."

"The number" as representation simultaneously holds and issues an appeal. It is discrete, it is easily subject to the algorithmic needs of models, and it communicates the authority of an imagined objectivity. But what the number appeals to is distinct from (though integrally related to) the appeal that the number holds. In the latter case, it attracts through its capacity to legitimate and to make actors and their interests, needs, and responsibilities visible, with all of the accordant gains that visibility generates. In explaining their articulation with TEEB at COP10, for example, modelers spoke of an opportunity for their models to have a policy impact; activists/ environmental groups saw an opportunity to use TEEB to reach policy makers and make them see "how the world really is," and ministers of environment sought an opportunity to demonstrate to ministers of finance that biodiversity does have a "value" figure that can be incorporated in national accounts. As one said, "In my budget I had 6 million pounds to address fungal diseases in honey bees. The Finance Department said get rid of it, and I said I could, but it would cost 190 million pounds. They asked why, and I explained the effect of fungal diseases on pollination and the cost of decline in yields, which had been quantified by our national accounts office. I got my 6 million pounds."[10]

Much of this appeal of the number is bound to the authority granted economics and cost-benefit analysis, but it is also related to shifts in the context of environmental decision making as practices of neoliberal governance have subordinated ecological rationales to economic rationales. Yet, these rationales also demonstrate that the intellectual apparatus behind the number with all its assumptions and calculations is incidental. Its power to convince is what really matters. As the head of UNEP Media reflected, "TEEB's gone from . . . a kind of interesting subject for environmental correspondents to one now where business correspondents and the politicians are getting the message. One [reason] of course is the numbers. Sheer numbers make one sit up in bed, don't they?"[11] Like technologies of visualization, such as maps, models, and narratives designed to make nature legible (Scott 1998) and visible (Brosius 2006; Forsyth 2003), numbers create nature as understandable and approachable for policy makers and thus mechanisms for remaking reality. By packaging a series of numbers, TEEB appealed to policy, business, and public audiences not only to support conservation but also to help *create* the conditions for the emergence of a market for ecosystem services.

It is the claims made on behalf of numbers and the sanctioning effect of those claims that give us insight into TEEB's primary claim, which is a moral one. TEEB leadership carefully crafted a message to seek win-win-win solutions that would simultaneously encompass the environment, the economy, and people. The constant refrain across TEEB sessions of nature being the "GDP of the poor" positioned TEEB, accounting, and the pricing of nature as projects that served the interests of "the poor." Sukhdev argued, "The central concern of TEEB is that the economic invisibility of nature has . . . exacerbated the suffering of human beings, especially those at the bottom of the economic pyramid. . . . That is the biggest finding that TEEB has to present to you today."[12]

In this explicit calculation, designed to appeal to development practitioners as well as conservationists, TEEB has become another moment in conservation's long struggle to become relevant to the poor. Like many such efforts, it endeavors to illustrate how, via its commodification, conservation can become compatible with poverty reduction (Büscher 2010a). This utility of "the poor" in the promotion of the financialization of biodiversity is instrumentalist at root. In terms of degradation, biodiversity loss is greatest in areas subject to industrialization and urbanization, and that in fact is where we find the majority of the world's poor—those without access to land and without access to clean water or air. It ignores the fierce and often violent battles over property and property rights that market mechanisms open up and appeals instead to social justice as a moral quantity best pursued and distributed through the market. In making nature visible and legible, the number abstracts and decontextualizes socionature and subsequently reembeds it in society (McAfee and Shapiro 2010), translating socioecological characteristics into a "nature" that capitalism "can see" (Robertson 2007).

The crucial moral appeal of TEEB, however, lies in implicit assumptions about rationalism and policy making. During COP10, Pavan Sukhdev stated: "Economics at the end of the day is the currency of policy, and it's important to get the economics right. But economics at the same time is only weaponry. The direction in which you shoot is an ethical choice."[13] The reliance of TEEB on rationalism for its own legitimation is readily apparent: "Understanding and capturing the value of ecosystems can lead to better informed . . . decisions; accounting for such value can result in better management; investing in natural capital can yield high returns; and sharing the benefits of these actions can deliver real benefits to those worst off in society" (TEEB 2009, 3). Sukhdev frequently repeated a phrase from management school texts: "What you do not measure, you do not manage."[14] Trite as this sounds, it is significant, since it frames the question of legibility, or the way in which a world (i.e., nature) comes into being through the production and accumulation of "facts" about that world.

These comments are grounded in a crucial assertion that "business" and, more problematically, government have not been acting rationally— that in allowing the degradation of ecosystems and biodiversity, they have been undermining the capacity to accumulate wealth. Yet, rather than see this problem as malignant—as a contradiction of capitalism—it is viewed as a function of not having the right "information." As such, these are also claims regarding the morality of metrics, as if to say that what is fixed quantitatively can be acted upon qualitatively—if policy makers had the right (quantitatively correct) information, they would make the right (qualitatively correct, i.e., moral) decisions—and that rational decisions cannot be made in the absence of "the right information."

If TEEB is the rationalist device meant to produce nature as capital, the success of that alignment is strongly attached to the qualities of those performing it, and rationalism requires the embodied enactment of expertise to legitimate its authority (Carr 2010). Carrier and West (2009, 7) acknowledge this when they point out that some agents are better placed than others to conform the world to a virtual vision: "The visionary must be powerful politically and the vision must be grounded in a form of knowledge production that is powerful socially." As the singular consistent embodied presence of TEEB, Pavan Sukhdev presented himself and was produced as a visionary for natural capital:

> If you want to ask when the first glimmerings happened, it was when a friend of my wife's asked me, "Why are some things worth money and other things not?" Economics treats . . . nature and its flows, its benefits, as externalities, and her question was very simple and very important. . . . I have kind of understood the issue, perhaps earlier than the average man on the street, and I just felt it was my duty to bring it out, to do as much work to develop this issue and understand why it is that we can't seem to account for what's valuable.[15]

These words position Sukhdev as a visionary. However, the production of a visionary also requires the sanction of other politically powerful actors. Where academic ecological economists failed to mobilize environmental institutions and organizations, Sukhdev has successfully directed the integration of their models into conservation institutions such as the CBD, conservation NGOs, states, and private-sector actors.

In many ways, the success of TEEB was tied to features that address the desire among CBD parties and other conservation organizations to engage with "nontraditional" actors. Sukhdev's credentials as a "conservation outsider" served to legitimate his expertise. As the UNEP media official stated, "The success of TEEB is [that] we have someone like Pavan who's available all the time for press, for media, for interviews to get the message out, with the credibility of being a banker, right? He wasn't from an environmental NGO, so he wasn't part of the converted, although of course he has been converted."[16]

As the TEEB visionary, the embodiment of expertise, recognized and sanctioned by a loose coalition of powerful actors, Sukhdev was able to help shift the CBD in a new direction. He reflects what Greenwood and Suddaby (2006) have termed "institutional entrepreneurs," actors who support institutions that promote interests the actors value but that have been previously suppressed by other actors or logics. During COP10, for example, as the prominence of TEEB became evident, side event titles changed, corridor conversations shifted, and high-level politicians struggled to reformulate their speeches in the language of ecosystem services and, more specifically, TEEB. Sukhdev frequently appeared on a variety of stages with other powerful actors, and his enactment of expertise and authority underpinned this capacity to achieve conformity—to enroll a wide range of actors across the event and beyond it, across networks that spanned private, nonprofit, and public sectors. Ultimately, TEEB cannot perform, cannot become part of the agencement, and cannot do the work of realizing the virtual reality of natural capital without the voice(s), like Sukhdev's, that lend it the sanction of expertise and authority, the stage(s) upon which they enact expertise, and the audiences for whom they perform. This is what makes virtual moments like TEEB and instances like COP10 integral to, and integral to understanding, the production and legitimation of "natural capital."

Conclusion

> So, as nature has changed in human eyes, the ways that we
> deal with nature and each other have changed as well.
> —JAMES G. CARRIER and PAIGE WEST, eds.,
> *Virtualism, Governance and Practice*

Bringing the world into being as natural capital is an ongoing and dynamic exercise in virtualism, where TEEB is a moment in the longue durée of the virtualism of natural capital. If virtualism is the process through which "reality" is made to conform to virtual reality, we describe a moment in the virtualism of natural capital and examine an emergent political project—TEEB—as one key step in conforming image to reality. However, describing the emergent implications of that moment requires a capacity to situate it as an agencement that maps the heterogeneous ensemble of actors, institutions, and devices (the apparatus) engaged in the production of natural capital and the dynamic and contested relations among them.

Carrier argues, "What distinguishes economic abstraction is the combination of its institutional power and its tendency to slip into virtualism. This is the conscious attempt to make the real world conform to the virtual image, justified by the claim that the failure of the real to conform to the ideal is a consequence merely of imperfections, but is a failure that itself has undesirable consequences" (Carrier and Miller 1998, 8). We do not disagree with this, but our analysis of TEEB suggests three modifications:

1. The institutional power Carrier highlights does not precede virtualism but is also brought into being as virtualism realizes some measure of "success,"
2. Virtualism is not something that is slipped into. The "slip" is a march — it is orchestrated, structured, scripted, and contested. Virtualism is achieved through performance that facilitates the reproduction of an agencement (i.e., the articulation and alignment of actors, institutions, devices, technologies, and methodologies) (Hardie and MacKenzie 2007).
3. Human actors know they are engaged in performance and acts of articulation and alignment. While virtualism begins with an ideological commitment, it must also be achieved through virtual moments that are linked together in an ongoing process of reproduction grounded in conditions of contestation. It relies on processes of alignment and articulation that draw powerful actors together to subsequently enact that virtual reality with an aim to establishing durable associations in ways that institutionalize and subsequently operationalize those models to convert abstractions into reality.

It is through rendering a valued nature "legible" (i.e., priced and costed) for key audiences that TEEB, as a component of natural capital, has been able to mobilize a critical mass of support ranging from modelers to policy makers, parliamentarians, and bankers. In its acts of reducing the complexity of ecological dynamics to idealized categories and in claiming to be a quantitative force for morality, TEEB is engaged in the production and circulation of practices designed to conform the "real" to the virtual. Understanding these acts of conformity, we argue, requires attending to the spaces where the performance of this model and the "facts" it produces are made apparent. The CBD is one such site where the discursive strategies through which TEEB mobilizes the alignments and articulations required to overcome obstacles to the realization of "natural capital" are readily apparent.

Indeed, it is these alignments and articulations that are a condition of TEEB's production. Contrary to what proponents would assert, it is the network of attached actors that is TEEB, not the substance. As we pointed out above, the ideas contained in the TEEB study are not new. What is new is the packaging, its attendance to specific audiences, the assemblage (institutional conditions) that contribute to its prominence (presence), and the capacity of those conditions and the presence they provide to draw actors to the package. TEEB is more than simply an ostensible product "for sale" (or, as Pavan Sukhdev frequently repeated, "a gift"), it is a packaged good, containing premises, assumptions, models, and predictions, that is intentionally networked and articulated with a broader group of actors and devices.

It would be an overstatement, however, to exaggerate the possibility of such calculation, for as projects like TEEB become dominant—as they are instutionalized—choice is constrained and articulation becomes more likely, especially if smaller actors seek to retain legitimacy and funding within the network of institutional environmental governance (MacDonald 2010b). As we witnessed during COP10, sanctioning TEEB as a core mechanism of the CBD is one way to lend it institutional coherence and to mobilize alignment and subsequent articulations. The ramifications of this are difficult to predict, but the analytic utility of witnessing TEEB being converted from study to tool is that it provides the ability to track its deployment across space and to make more sense of the relations involved in its circulation and both the policy and material ecological effects of that deployment. Since the rollout of TEEB at COP10, for example, it has rapidly circulated through subsequent meetings related to biodiversity. A case in point was the January 2011 Symposium on Caribbean Marine Protected Areas, held in Guadeloupe, where a representative of Fonds français pour l'environnement mondial, Paris (the French focal point for the GEF), referred to "Nagoya, where a major event was the publication of a study of the valuation of ecosystems [TEEB], made public during the convention," and the moderator introduced TEEB to the assembled audience of protected area managers, academics, and state and NGO representatives as "the international bible of socioeconomic assessment."[17] The biblical status that TEEB seems to have earned so quickly reflects the shifting ideological and material landscape of biodiversity conservation, where a new "reality"—a new ontology—is being brought into being by reordering relations of power around the ideological project of "accounting for nature" and the political project of convincing business and policy-makers that nature is valuable because it can be priced (see also Mac-Donald 2010a).

While we have restricted much of our analysis to a particular project in the production of natural capital, it is important to highlight the relations between the processes of alignment and articulation that we have described here and how they reflect the containment of an effective oppositional politics and the very possibility of imagining natural capital. In a volume on virtualism in conservation projects, Carrier and West (2009, 1) recognize environmentalisms as different kinds of "ways of thinking" that "intersect with the world and people in it" and, consequently, the ways in which people identify and evaluate their natural surroundings, but they give fleeting mention of the ways in which environmentalism has become a vehicle that operates in the interests of capital accumulation and a vessel to be claimed in the legitimation of distinct projects. Once seen as a singular and distinct threat to accumulation, "environmentalism" has become in practice a politics that can be enlisted, contained, and directed to the interests of capital accumulation.

TEEB is indicative of this process. Its rhetoric of crisis and value underpins a larger political project that aligns capitalism with a new kind of ecological modernization in which "the market" and market devices serve as key mechanisms in practical efforts to conform the real and the virtual. The consequences of this, however, are material and have been expressed by others who have described how the ascendance of neoliberal conservation has shifted the locus of decision making in international conservation (Corson 2010; MacDonald 2010b). Nowhere is this more apparent than in arenas of international conventions, where states are granting their authority not just to private investors but to speculators who, desperate for a new and profitable investment frontier, are sinking their capital into the promise of nature and speculating on its scarcity, all the while describing their actions as environmentalism (Sullivan 2011a). TEEB is a step in this process, legitimating the market as the means through which biodiversity is conceived, stabilized, and exchanged; it is the realm in which economic rationale, in realizing new forms of accumulation, displaces ecological rationale. Within this realm the financial modeling of nature provides critical new investment opportunities, and the construction of environmental services as commodities opens them up to speculative behavior, as calls for internalizing environmental externalities are transformed into the "optimistic embrace of the returns that might be captured if this 'value' of environmental externalities can be priced and traded" (Sullivan 2011a, 7). We argue that TEEB is playing an important role in legitimating and circulating the narratives, images, and ideas of nature essential to these new speculative nature markets.

As projects like TEEB become instruments for capital expansion, they become agents of nature's restructuring, underpinning what Bram Büscher (this volume) calls "one of the biggest contradictions of our times": the idea that "nature can be conserved by increasing the intensity, reach, and depth of capital circulation." That contradiction is the virtualism of natural capital. Increasingly, modes of conforming reality with the image of natural capital circulate in popular culture and the daily economy of life. As travelers purchase carbon offsets to assuage the "guilt" of flight and as schoolchildren come to understand trees first and foremost as services in the reproduction of capital accumulation, we move closer to the virtualism of natural capital.

Notes

The authors would like to thank participants in the CBD Collaborative Event Ethnography (CEE), a group of researchers engaged in the ethnographic study of conservation institutions. CBD-CEE members conducted collective fieldwork during the 2010 Conference of the Parties to the Convention on Biological Diversity and are committed to the collective sharing of data and analysis arising from that fieldwork.

1. Bernard Payne, question put to Al Gore in the *Independent*, July 7, 2007; response to Bernard Payne from Al Gore in the *Independent*, July 7, 2007. Note that Lohmann (2009) also cites this quote by Al Gore.

2. Sukhdev also chairs the World Economic Forum's Global Agenda Council on Biodiversity and was a speaker at Davos in 2010 and 2011. He serves on the boards of Conservation International and the Stockholm Resilience Centre.

3. Hall defines "articulation" as both the joining together of diverse elements and the expression of meaning through language. For Hall, articulations are made in historically specific contexts. They are temporary, contingent on material and discursive factors, but never determined. We use articulation to refer to both the realization of that affiliation through the linking of their ongoing activities with the agencement and the publication of that linkage (aurally, textually, and visually) in ways that extend the presence and contribute to the strength of the agencement.

4. Performance includes the scripting, structuring, and staging of economic expertise. By alignment we mean the orientation of actors toward the virtual reality out of which the virtualism is created, in our case, natural capital. Alignment is facilitated through both legibility and visibility, when actors are exposed to the configuration of power around what they see as particular projects and envision ways in which their multiple interests (personal, organizational, institutional, etc.) can be met through an affiliation with the agencement.

5. Based on *The Stern Review on the Economics of Climate Change*, a study led by Sir Nicholas Stern, head of the UK Government Economic Service and advisor to the government on the economics of climate change and development.

6. Additional members of the market group included Dan Suarez, Shannon Greenberg, and Juan Luis Dammert Bello.

7. SBSTTA 14, -XIV/4 (c) 8, http://www.cbd.int/recommendation/sbstta/default.shtml?id=12251.

8. UNEP/CBD/WG-RI/3/L.9, "Updating and Revision of the Strategic Plan for the Post-2010 Period," May 28, 2010, http://www.cbd.int/sp/notifications/.

9. TEEB press conference, 10th Conference of the Parties to the Convention on Biological Diversity, Nagoya, Japan, October 20, 2011.

10. "GLOBE International: Legislative Approaches to Recognizing the Value of Biodiversity and Natural Capital," COP10 side event, October 27, 2010.

11. "TEEB 4 Me: Communicating the Value of Nature," Ecosystem Pavilion, COP10 side event, October 25, 2010.

12. TEEB press conference.

13. Ibid.

14. This phrase has been repeated so frequently by Sukhdev that, despite the fact that it has been circulating for decades, some people in the conservation world have begun to attribute it to him.

15. "Dr. Pavan Sukhdev on the Invisible Economy," YouTube Channel of Corporate Knights: The Canadian Magazine for Responsible Business, http://www.youtube.com/watch?v=VZWnMaX_bsY. Sukhdev is widely distributed in video as the face of TEEB.

16. Nick Nuttal, TEEB 4 Me.

17. We are grateful to Noella Gray, University of Guelph, for allowing us to use her notes of this encounter.

CHAPTER THREE

Nature, Villagers, and the State

Resistance Politics from Protected Areas in Zimbabwe

Frank Matose

This chapter addresses what Harvey calls "accumulation through disposses-sion" (2003, 93) insofar as conservation leads to dispossession, alienation, and ultimately impoverishment of rural people around protected areas. Through a case study of protected forests in Zimbabwe, the chapter argues that the colonial and postcolonial Zimbabwean state implemented prac-tices to dispossess and alienate forest residents from protected forests to make way for commercial timber harvesting and wildlife hunting. During colonial times but also today, this was a deliberate strategy that should be seen in the context of forcing local people to become part of the reserve labor pool for (developing) settler capital. However, over time, forest residents developed various strategies of resisting the state bureaucracy's attempts at dispossession.

Many studies have explored resistance around forest reserves (e.g., Guha 1989; Peluso 1992; Moore 2005), following on James Scott's studies in Southeast Asia of this phenomenon. This chapter traces how values placed on forest resources by different actors (see Matose 2002), institutional arrangements, and the way policies are framed in different contextual set-tings conjoin to construct particular types of relationships among different sets of people around protected forests. The interaction of domination and struggle results in a variety of outcomes around forest reserves (Haynes and Prakash 1991), characterized by conflict, resistance, and different forms of collaboration or cooperation. Different actors resist the impact of policy framing and institutional arrangements in different ways, yet this is not always easy to recognize, as for many actors "hidden transcripts" rarely come to the surface (Scott 1990). Hence, Scott (1985) earlier argued that studying the "weapons of the weak"—that is, the subtle struggles waged by the less powerful, such as villagers—is very revealing; the otherwise mundane activities of the less powerful become the more pertinent to study, as they represent the everyday struggles against more powerful state and capital agents. Thus, in this chapter, the emphasis is placed on agency *within* rather than on analyzing the broader structural political economy *of* neoliberal conservation, upon which many other chapters focus. Communities that live within and around reserved forests in Zimbabwe are not known for their open confrontations with the Zimbabwe Forestry Commission (FC). However, latent forms of struggle not only are waged against the FC's forms of power and hegemonic control over resources but also involve a variety of struggles *within* heterogeneous communities.

This chapter focuses on the agency of local people by engaging the quotidian politics and the moral economy of African villagers through a discussion of power, domination, and resistance vis-à-vis the state's neoliberal accumulation project around protected areas (see Scott 1985, 1990; Hanchard 2006; Holmes 2007). The chapter adds to recent scholarship around the critique of conservation (see, e.g., Brockington, Duffy, and Igoe 2008; Büscher 2009; and chapters in this volume, among others) by exploring how local people negotiate state dispossession and exercise their power based on Foucault's understanding of power and Giddens's (1984) ideas about structuration. Foucault (1982, 217) states that power is in many formulations an "all-embracing and reifying term" such that in order to understand it more meaningfully, what becomes pertinent is to study the multiple power relations in a given context. Struggles over power relations are universal; they are found in multiple localities, transcending social and economic boundaries. Such struggles are over the effects of power, but they are also directed at the framing of knowledge about forest environments.

In the ensuing discussion, then, power is analyzed in terms of the way forest policies have been framed around forest reserves in Zimbabwe, particularly since the 1990s, as will be discussed in section 1. Power relations emanate either from application of capacities or through discursive practices, a process that entails both knowledge politics and direct overt domination. Power relations have to be understood from the perspective of their origin, their nature, and how they become manifest. Foucault (1982, 219) argues that "power exists only when it is put into action." The way forest policies are framed is therefore a reflection of the exercise of power by forestry authorities. Nonetheless, forest policies do change, and understanding this requires an understanding of the concept of agency. For this understanding, the discussion turns to Giddens's theory of structuration.

Giddens (1984, 25) defines structuration as the "conditions governing the continuity or transmutation of structures, and therefore the reproduction of social systems." The structures are the rules and resources that facilitate the production and reproduction of social systems over time and space. Analyzing the structuration of social systems "means studying the modes in which such systems, grounded in the knowledgeable activities of situated actors who draw upon rules and resources in the diversity of action contexts, are produced and reproduced in interaction" (Giddens 1984, 25). This means that people are knowledgeable agents who are actively constituted by and reconstitute social systems from their social interactions. Change is effected by people's agency. Giddens defines agency not in terms of people's intentions in doing things but in terms of "their capability of doing those things in the first place" (9).

Building on this work, attention is paid to the informal contestations of domination by the state by different social groupings. I argue, following Scott (1985, 1990), that the "weak" — in this case, villagers — use both overt and covert forms of resistance. An attempt will be made to deepen discussion of whether the consequences of dispossession and neoliberal conservation for the weak are often more indirect, disparate, and individual than collective and overt and whether, in relation to this, one can discern different categories and types of resistance. More concretely, the chapter picks up on three threads from critical conservation literature. The first of these is that the phenomenon of the "scaling back of the state" associated with neoliberalization (Igoe and Brockington 2007, 436) does not mean that the state disappears; instead, it is merely reformed (see, e.g., Büscher 2010c; Fletcher 2010). Since the start of the 1990s, the structural adjustment imperatives of the World Bank have been foisted upon the Zimbabwe state, resulting in cutbacks of state funding to conservation, in turn providing impetus for the FC (as a state corporation) to raise this funding gap from the protection of state forests. (It has to be noted that since 2000 Zimbabwe has suffered a massive economic and social downturn due to political turmoil in the country that has only somewhat abated since 2009.) This issue will be returned to in the next section.

Allied to the first point, neoliberalization has also led to the prolifer-ation of privatization in order to make conservation a success (Igoe and Brockington 2007). In the case of Zimbabwean forests, state collusion with private capital has always been a feature of green capitalism since the turn of the twentieth century (see, e.g., Kwashirai 2009), but since the 1990s, the marketing of forests for funding conservation has gained new impetus, as discussed further below. The third focal point is in relation to what Igoe and Brockington (2007, 437) refer to as "territorialisation"—that is, the offer of state-controlled territories to investors through rents and concessions, an offer, as the authors point out, that predates neoliberalization but has become more pronounced because of it. In the case that will be discussed, territorialization takes the form of safari hunting and timber concessions to private capital. All of this has taken place under the cloud of a "discur-sive blur" (Büscher and Dressler 2007) in the sense that increasingly I witnessed the forests being protected under the policy discourse of "biodi-versity conservation" in sync with discourses elsewhere in the world but in contradiction to local realities and practices (Matose 2002, forthcoming).

The chapter is structured as follows. Sections 1–3 provide the backdrop to the overall argument by describing the setting of the protected forests. Sections 4–6 describe what local people do in different contexts to express their power/agency. In section 7 the chapter returns to the main argument about the effects of the state's dispossession agenda on local people.

1. Background to the Zimbabwe Forestry Landscape

About 2.4 percent (9,414 square kilometers) of Zimbabwe's land area is forested and managed on behalf of the state by the FC, which operates as a quasi-autonomous agency that is both focused on conservation and a capital-generating, business-like entity. A study site from one of the pro-tected forest/state reserves located in the western part of the country was selected for this research, conducted mostly between 1997 and 1999, with less intensive visits in 2002 and 2011. The case study area included the Gwayi/Mbembesi Forests (hereafter Gwayi for convenience) located in the Lupane and Bubi Districts of Matebeleland North Province. To the south, the forest shares the border with Umguza District, while to the south-west lies Tsholotsho District. There are resident communities on borders with the other districts that use forest resources, making the rural district councils (RDCs) part of the overall forest-outreach system. Each district is subdivided into several wards. It is from two such wards that the two villages for the study in the Gwayi case were selected. This case itself was selected on the basis of being the biggest and oldest forest area, at over 180,000 hectares (combined area), and gazetted in 1931 and 1940, respectively, for each part of the forest. The forest has a long history of settlement by largely Nbebele-speaking people, stretching back before state dispossession during colonization (1900–1940) (see Kwashirai 2009). The rest of this section highlights how forestry conservation in Zimbabwe has evolved in the framework of neoliberal pressures.

First, the 1990s witnessed the growth of the tourism industry in Zimbabwe as the neoliberal economic dictates of structural adjustment programs (SAPs) began to take effect and led to the scaling back of the state (Ferguson 2006; Igoe and Brockington 2007). Whereas the FC prior to the 1990s received full state funding for conserving forests, after SAP implementation less and less funding came from the central state, with this gap being filled by revenue derived from the new opportunities created by the neoliberal environment of foreign exchange deregulation. Tourism rose substantially as a result, bringing revenue from wildlife conservation, even within reserved forests like Gwayi. According to Hodgson (1989, 2), "the Forestry Commission has identified the Gwayi Forest as being particularly valuable as a wildlife resource, as it contains the complete range of ecosystems and an existing valuable wildlife population. The presence of the tenant farming programme, isolating the major riverine vegetation area, interferes with the wildlife programme. . . . It is this area that the Forestry Commission has identified as being particularly critical as a wildlife habitat."

A focus on revenue generation from wildlife utilization also increased the need to move residents from western Gwayi to eastern Gwayi in order to create another wildlife habitat following the success of the Insuza Wetland displacement in the 1960s, when forest residents were forcibly moved to their current location in western Gwayi (see Matose 2002). However, because of people's resistance, a policy of "harassment" was initiated to induce residents to move: "On the issue of water provision, he [the senior minister of local government and urban and rural development] said he had noticed, when he was Minister of Natural Resources, that animals get much better attention. He cited Hwange National Park as an example, where he said there is a water-point every 2km interval that runs all day, to make sure water spills. But the Forestry Commission is denying human beings access to water for drinking, restricting people to three times a week, he said it was ridiculous."[1]

Such relocation from the forests has been justified on two main grounds. First, the forest authorities argue that if any people are allowed to stay,

pressure will continue for others to settle, and this will lead to "the destruction of forests." Second, the FC argues that it is not in a position to provide
the level of infrastructure that is available to communal land residents
and therefore would be guilty of depriving forest occupants of the right to
share the development being enjoyed by those living in communal lands
by allowing them to stay in the forests (Matose 2002). The need to move
forest residents is often supported by "scientific" arguments such as the
following: "The actual and prospective environmental changes associated
with the settling of people in Gwayi/Bembesi forest areas is reflected by the
results. . . . Deforestation is the first in the chain of environmental problems
to occur. Removal of vegetative cover, as is happening in the settled area,
reduces nutrient cycling in the ecosystem, exposing the soil to heavy rain,
generating soil erosion and reducing soil moisture" (Mhuriro 1996, 32–33).

Forcing forest residents to move from landscapes that are identified for
wildlife habitat was also attempted but defied, as discussed in greater detail
below (see also Matose 2002). The attempts to remove forest residents
from the heart of Gwayi forests constitutes the second form of neoliberal
conservation practice. The claiming of territory (after Igoe and Brockington 2007) occupied by forest residents was meant to create space for
more wildlife habitat along the Mbembesi River (Matose 2002), but in
practice the idea was to increase the revenue streams that would accrue
from increased safari-hunting concessions. At the same time, the quote
above also illustrates the "discursive blur" that Büscher and Dressler (2007)
discuss in relation to neoliberal conservation (described in greater detail
in Matose, forthcoming). Since the 1990s and with the rise in significance
of the Convention on Biological Diversity (CBD), the FC (a signatory to
the CBD as well as implementing its provisions) has used international
discourses around this convention to justify practices that effectively marginalize forest residents. Table 3.1 depicts the number of villagers who are
being affected by the FC's need to depopulate the western part of the forest:
over 20,000 people occupy a mere 11,000-hectare portion of over 180,000
hectares of forestland.

2. Timber Concessions

Forest areas were originally set aside with the primary purpose of ensuring
a sustained supply of hardwood timber to a burgeoning settler industry
in the early twentieth century (McGregor 1991; Alexander, McGregor,
and Ranger 2000; Mapedza 2007; Kwashirai 2009; Matose, forthcoming).

Table 3.1. Overview of the populations around Gwayi Forest (2010)

Forest	Area (ha)	Year of entry	No. of house-holds	Popula-tion of residents	No. of cattle	No. of goats	No. of donkeys	Other livestock	Area settled (ha)
Gwayi	144,230	2004–5	1,124	11,012	5,136	2,726	9,052	1,012	6,000
Bembesi	55,100	1994–2001	928	10,444	3,964	2,328	1,055	1,121	5,200

Source: Forestry Commission Report 2010.

Revenue from timber concessions continues to be important for the FC in the Gwayi case study at both organizational and individual levels. As an example, in 1998 forest managers covertly sold timber worth over ZW\$1 million (US\$56,818; US\$1 = ZW\$18) to a saw-milling company, and each of the officers involved personally received around ZW\$200,000 (US\$11,363) in return. However, their covert rent-seeking practices were eventually discovered, and they were dismissed from FC service, although the money was not recovered (Matose 2002). At the organizational level, timber sales generated the highest revenue for Gwayi forest in 1998 (ZW\$2,677,000, or US\$148,722), accounting for 63 percent of the total income for the forest. Despite this, timber concessions do not necessarily realize the best values of the commercial timber due to weaknesses in the bidding process between the FC and the logging companies (Mushove 1993; Dore 1999). At the same time, timber revenue is not assured every year, as cutting cycles vary by species.

3. Wildlife Safari Hunting

For the Gwayi case under discussion, wild animals are a prominent part of protected forests in Matebeleland by virtue of being in close proximity to national parks, especially Hwange National Park. Since the 1950s, a thriving safari-hunting industry has been operating within protected forests, in part to fund the FC and in part as a component of a broader conservation strategy of the state. This is achieved through spreading the wildlife resource across different agencies. Due to the importance of the wildlife-hunting industry, in the 1960s forest tenants were forcibly relocated from the Insuza Vlei (wetland) to make way for a wildlife habitat and increase the capital derived from this activity (Matose 2002; see also Kwashirai 2009). Once forest tenants were forcibly moved, wildlife species were introduced into the re-created space, and over time the species diversity has increased in relation to the market (hunters') demand. The latest figures that could be obtained from the FC annual reports indicate that wildlife safari hunting generates over 60 percent of the agency's revenue as of 2009 (over US$500,000), now exceeding timber revenue.

These two sources of revenue have illustrated the increasing importance given to forest conservation for funding a privatized arm of the state that used to be solely supported from the Zimbabwe treasury prior to 1990. With the neoliberalization of the state since 1990, the revenues derived from conservation have increased in significance, leading to changes in practices that in turn have adversely affected relations with forest resident communities. In the next section, an analysis is made of why conflicts take place in certain circumstances, while resistance and other relations dominate in others in response to the state's dispossession and accumulation tactics.

4. Different Forms of Resistance around Protected Forests

Why do some conflicts become overt while others are manifest in other ways? First, some forms and types of resistance can be traced partly to the issue of cultural identity of the Ndebele residents (Matose 2002). The fertile valleys of the Kalahari Sand Dunes have been the Ndebele residents' home for many generations. This way of life came under increasing threat from the cumulative effect of the state's neoliberal conservation practices around protected forests, practices that threaten Ndebele residents' existence and culture. Cultural identity also revolves around keeping large livestock herds, which require access to good rangelands, and western Gwayi provides that. Grazing is a highly valued resource for older Ndebele forest residents in both Gwayi sites, and these residents' livelihoods are threatened if state regulations are highly enforced. These livestock owners would be left with no option but to take their battles into the open (see Matose 2002).

At the core of the conflicts are power-based negotiation over meanings and perceptions of what the Gwayi forest represents. Forest officials within the forestry bureaucracy at various levels formed an alliance to move residents from the west to the less productive east in order to pursue conservation objectives and to realize more value from an increased wildlife habitat. This move to relocate, thereby dispossessing residents of their land, was intended to raise revenues to cater to the dried-up state funding (the rolling back of the state) and also the commercialization and de facto "privatization" of the FC that took place in 2000. This had a precedent in the nearby Insuza Wetland in the 1960s, when wildlife was introduced in large numbers and became a source of revenue (Dore 1999). Older forest residents who own livestock resist relocation because of the great importance they attach to their way of life from livestock raising and farming along the fertile Mbembesi Valley, in contrast to the infertile sands on the eastern part of the forest.

What does this imply in relation to the arguments at the outset concerning the effects of neoliberal conservation on local people who in turn exercise agency as theorized by Scott (1985, 1990)? Hanchard (2006, 35) states that actors who resist "are coagulants in the sense that they infuse a relatively self-contained instance with their own notions of justice, equality, and redress to significantly affect micropolitical—and sometimes macropolitical—outcomes in daily life." This occurs through "episodic circumstances within their immediate environment" (33), and during these episodes actors engage in activities to express their agency in a variety of ways. What matters is the scale of protest and the degree to which dissent is articulated. Gross injustice leads to articulation at a variety of places, albeit at the individual level, but the cumulative effect is to be highly visible to the state, which is the intention of those overtly resisting the collusion of the state with capital.

Second, whether these different visions (transcripts) remain hidden or become public is linked to the state's pursuit of what Scott (1998) refers to as the "high modernism agenda," in this case, the ordering of nature through the practice of forestry. Sustained yield management illustrates Scott's framing by organizing landscapes into forest reserves in the early twentieth century in the form of blocks within each forest from which periodic harvests are organized. The ordering of nature obviously affects forest residents, compounded by the privatization or intense commercialization of forest resources by the neoliberal state since the 1990s. Prior to that, when forest conservation was fully funded by the state, the FC allowed forest residents access to game meat and living off Gwayi Forest. Elsewhere I trace how these practices have been put into effect since settler colonialism at the beginning of the last century (Matose, forthcoming).

As a result of the state's long-established practices, villagers/forest residents have developed long-term strategies to assert their agency. These assertions of power take the form of "hidden transcripts" (Scott 1990) that are a way of life for most forest villagers. Their way of life has evolved around a technical bureaucracy that attempts to tame nature as an accumulation strategy (Smith 2007), with villagers forming an "inconvenient" reserve labor pool. Forest residents who in the early part of the twentieth century were convenient to the forestry enterprise as reserve laborers are now considered inconvenient to a much more robust commercial conservation, while their population has risen. State attempts at social engineering (Scott 1998; more robust details in Matose 2002) have led to the development of parallel cultural practices that are hidden from the state's gaze but highly intentional. Nonetheless, the effects of these practices on the forest landscape would be visible to state forestry officials, even though the acts themselves might have been hidden.

Any of the neoliberal conservation practices referred to in the previous section may lead to forest residents engaging in such forms of covert resistance. Scott (1985, xvi) describes these acts as "the ordinary weapons of relatively powerless groups . . . false compliance, pilfering, feigned ignorance, slander, arson, sabotage . . . [that] typically avoid any direct, symbolic confrontation with authority." Such acts are intentional on the part of the villagers/actors, and they are often unnoticed, although in my experience, the effects of these acts were often noticed. Acts of incendiarism would always have a profound effect on the forest landscape and would be noticed by the targeted state forestry officials. In many cases, those performing such acts would go unpunished, as they would be very difficult to apprehend. The cumulative effect of such acts, no matter how mundane, would impact the state's policies regarding forest villagers/residents. The rest of this chapter documents the agency of forest residents in relation to state conservation practices in terms of the framework provided here.

5. Overt Resistance

Case Study 1. Land and Pole Tickets in Gwayi West, September 1997

Things came to a head in terms of relations between the FC and communities in the west to the extent that the forester in charge was beaten by one youth from the community. My research work was suspended for a while due to tempers that were running so high in the area that it was too dangerous to have discussions with residents. On September 16, 1997, the forester for Gwayi was driving his truck within one of the residents' villages. It had rained the previous day, and there were pools of water in the dirt road. When the forester approached one pool of water, instead of avoiding it by driving around it, he drove through it and splashed a young man who was walking along the same road. The young man was so infuriated (not least because of the fines that his mates had been issued by this same forester) that he whistled for the forester to stop. The young man then walked over to the driver's side of the car and slapped the driver, daring him to come out of the car. The forester asked the young man why he had slapped him. The young man replied that the forester had gone too far, and if he carried on giving residents problems, his life was going to be in danger. The forester then drove off, heading for the safety of the administrative offices. When I talked to the head of the Residents Association (RA) about the incident, he replied: "Of course, these young men react differently to being harassed by forest authorities. Sometimes that's what these rude foresters need, to be beaten once in a while. The young man had been denied the opportunity to clear land for farming, and then he is splashed with muddy water, while at the same time these foresters don't give you a lift. This is too much!"[2]

Rarely did contestation develop into open conflict around protected forests. However, between 1996 and 1997, relations between forest managers and forest residents in Gwayi West became so poor that they reached a boiling point. Overt conflict emerged over the use of land in Gwayi West, pitting forest residents' tenure against the enlargement of wildlife habitat for forest managers. Only in the incident above was there a physical exchange between a resident and the forester for Gwayi. Even the mechanisms used by different managers during the period in question led to forest residents describing the conditions described in case study 1 as "warlike" because forest residents reminded managers of the 1970s liberation war, which led to Zimbabwe's independence in 1980. Such mechanisms included the closure of water wells and dip tanks, leaving residents to rely on sand wells dug in a nearby river for their water needs and to buy chemicals for livestock dosages. These mechanisms were part of the general practice of "harassment" to which the FC resorted in order to compel Gwayi West residents to move to Gwayi East after initial refusal in 1995. The coercion of residents was all part of the neoliberal conservation imperative that the FC developed post-SAP, as the rolling back of the state led to attempts to increase territory under direct FC control—in other words, "territorialization"—in order to meet survival needs in the absence of state support. At the same time, eviction fitted into the overall scheme of commercialization in which poor, nonpaying forest residents had no role. In this way, the FC operates as a private enterprise when in reality it is a state conservation arm.

As a result of the pursuit of neoliberal conservation policy, there were increased attempts between 1993 and 1998 to relocate all forest residents from the west of Gwayi forest to the east. In order to realize this aim, forest managers resorted to the prohibition of expansion of settlement and cultivation areas for all forest residents. This marked the beginning of heightened conflicts between the two groups. Not only did residents in Gwayi West resist the relocation, but they lobbied various political and government hierarchies to assert their right to stay on forestland. Conflict was shaped by the residents' long history of movement within the forest. For example, some had to make way for wildlife in the Insuza Wetland in the late 1960s, and further movement was strongly resisted (Matose 2002). This was not least because there is a "black government" in power and "black" administrators running forestry affairs from whom forest residents expected more support than from colonial officers.[3]

Conflicts were fueled further by the FC's lack of consideration of basic social issues for which the organization had been responsible prior to the neoliberalization of the state in the 1990s. The FC used to provide several services to forest residents with funding from the central state, which was then terminated. At the peak of the 1997–98 period, marked by frustrations on the part of forestry officials concerning the failure to realize the territorial expansion for wildlife, forest residents were subject to "harassment." Not only were dip tanks and water wells closed, thereby cutting off livestock and forest residents from important sources of sustenance, but school operations were undermined through the prohibition of school reconstruction. School buildings in Gwayi West were mostly brick under thatch for classrooms, while teachers' dwellings were constructed from poles and mud.

During this period, Gwayi West residents lived in perpetual uncertainty and anxiety about their future within the forest, especially given the experiences of 1990, when some homes were burned and some people moved to distant communal lands (see Matose 2002). Foresters and forest managers also continuously told residents to leave. There was no longer any reference to the previous permit arrangements, which had given residents the authority to reside on forestland at least up to 1987. In some parts of Gwayi West, the FC had granted grazing leases that effectively reduced the amount of land available to local residents in certain villages as a means to compel them to move east. Grazing leases were part of the privatization/commercialization of forest resources under the neoliberal conservation culture that had enveloped the FC in the 1990s. Forest managers (the same ones who were subsequently fired for selling timber fraudulently) were alleged to be "very harsh and hostile as well as brutal."[4] Frequently, the denial of permits for construction materials was accompanied by threats of eviction from the forest altogether. Residents who resisted such permit denials or whose homes were in such a state of disrepair that they went ahead and cut poles and thatch grass without permits received irregular fines from the FPU (Forest Protection Unit, an armed wing of the FC charged with antipoaching activities). These fines ranged from as little as ZW$200 (US$11) to as much as ZW$4,000 (US$222), but most of the residents charged with such fines could not afford them. This was followed by the destruction of the newly established homes of grown-up children.

The residents' representatives eventually sought mediation by the governor of the province (as a senior ruling party official and the highest government representative) in the deteriorating situation.[5] As the governor described the circumstances in a separate meeting much later, "the forestry officials use archaic legislation to repress fellow black people."[6] This was in reference to the unchanged Forest Act, which failed to reflect forest dwellers' current needs and circumstances. The governor was then instrumental in getting the FC board to facilitate the ending of residents' "harassment" by managers. As a result, permits were resumed pending the outcome of negotiations for a potential Shared Forest Management Project. At the same time an FC board commission was instituted to review policies concerning forest residents, and this in turn ordered officials to suspend their harassment until permanent solutions for the residents could be found. As of 2011, such permanent solutions were not yet in place.

Thus, in this first type of resistance, the cumulative effects of neoliberal conservation practices led forest residents to exercise their agency in the open, but this in turn led the state to develop even tougher strategies to effect its policies, resulting in an interplay between structure (the state) and agency (local people's practices) in these power relations (Giddens 1984). Attempts to move forest residents from Gwayi West, pursued to dispossess local people from their ancestral lands in order for the FC to accumulate more value through the sale of wildlife, were successfully resisted by residents who continued to live on their lands within the protected forest.

6. Covert Resistance

Hidden transcripts (Scott 1990) emerged from the state practice of privatizing or commercializing forest resources, in terms of which forest residents were marginalized from resources they had previously accessed through a permit system. This prompted reactions different from those detailed in the previous section. The most pervasive form of everyday resistance—pilfering or theft—was witnessed in relation to the practices of game hunting, as illustrated in case study 2, below. Hunting occurred in both villages but was more pronounced in Gwayi West, where there were more long-term inhabitants, who have more of a history of hunting than more recent immigrants to forest areas. There was also less hunting in Gwayi East because of the highway barrier and the veterinary fences, as well as the distance to the Insuza Wetland. Hunting is as much about asserting rights and identity as securing a protein source. For hunters to be successful they have to operate in groups that endure over time and to be able to stay ahead of the FPU. At certain times hunters also rely on the wider community not to "sell them out" to forest authorities after hunters' tracks lead into a village. There would therefore be group resistance in the sense that a whole village would be silent about the identity of a hunter or hunters when residents knew full well who had been in the forest recently. Hence a hide-and-seek game ensued that sometimes forced FPU members to be stern and search every homestead for game meat.

Case Study 2. Game Hunting

The following is an extract of an interview with a young man in Gwayi West about hunting:

> We usually arrange to meet at some place inside the forest close to where one of us would have sighted a reasonable herd of eland. This will be my brother, our next-door neighbor [who actually lived a kilometer away], and a cousin and two of his friends. Our neighbor and cousin who is friends with some of the forest guards would have found out where they would be patrolling in the next few days in order for us to have a better chance of not being caught. We then take our dogs to the arranged meeting place around sunset. Hunting is usually done when there is full moonlight so that we can see at night. An eland herd with calves is the one targeted for easiness of kill. Dogs then chase the animals until they are tired. We will be running behind to find them in a circle with their heads in the center and kicking outwardly at the dogs. This is what eland do when they are tired. We then move closer and stab at those we can. At most we usually get two before they run off again. We then skin the animals very quickly and hide the skins somewhere and allow the carcasses to be drained of blood before setting off back home. The meat will have already been divided into equal portions for each person. We then ferry the meat in bags. We usually approach the village from indirect approaches from the forest to avoid our tracks being easily detected by the forest guards.[7]

Because of the increased presence of members of the FPU, hunters no longer waited for the meat to dry while still in the forest. If the team were unsuccessful in one night, they would return home rather than sleep out and risk the chances of being caught.

Case Study 3: Grazing Conflict in Gwayi East

As demonstrated by the following case study, conflicts over grazing areas were witnessed not only in Gwayi West but also in Gwayi East:

> I was a little mischievous, my children, and I cut the vet [veterinary] fence across the Falls road [Victoria Falls to Bulawayo Road] and let in my starving cattle. I had to save them from dying, as I had lost three already to this drought. As you can see, there is no grazing on this side at this time of the year, and I couldn't watch my cattle die daily. I asked the forestry officials at Forest Hill [20 kilometers away] to help us with the grazing area that is not leased across the road, but they sent me to the vet people, who in turn referred me back to the foresters. I came back with no clear answer. At the same time the lessee of the grazing area directly across from us waters his animals in our river, in the green zone every day, and takes them back to the red zone. However, when I asked to do the same I was thrown back and forth. I decided therefore to be mischievous and cut the fence to let my cattle survive. Now the vet people together with foresters are coming to shoot my cattle for crossing into the red zone and back into the green zone illegally, as well as for cutting their fence. Now, if my cattle die, then my children will not know what cattle looked like, and thus my family will be dead. That is why I have called all the village men to come and witness the shooting of my cattle and my death. I have had to ask for permission from the school in order for me to attend this case.[8]

The narrator of this story was a rich old resident of Gwayi East to whom grazing was very important for his household welfare. He even risked his life to save his cattle. The case demonstrates the inherent conflicts between the FC and cattle owners over grazing resources in Gwayi East. Relocation, together with the enforcement of veterinary regulations, would reduce available grazing in the eastern part of the forest. Large cattle owners, who are mostly earlier inhabitants, are unhappy with the FC, and as the case amply demonstrates, conflicts can become open at any time, given the covert operations to ensure that livestock have access to better range in the west.

These cases highlight hidden transcripts that emerge from the state's neoliberal ordering of nature (Scott 1998), which resulted in the privatization of resources such as game meat and grazing areas that excluded forest residents. The two cases of hunting and grazing represent people exercising their agency in a recursive way (Giddens 1984) in the face of constraints imposed by the state's domination of nature. In both cases, forest residents reflexively decided to take actions against the structural conditions imposed by the state.

Resistance also took the form of more mundane actions in reaction to any particular neoliberal conservation policy and practice of the FC. Many forest-harvesting activities that were permitted prior to the mid-1990s became criminalized with the commercialization of the FC. However, residents employed several covert strategies to retain access to forest resources.

Defiance is one form of resistance witnessed in Gwayi West. Women continued to harvest thatch grass when permits were unilaterally frozen—and collection prohibited altogether—during the 1997 season. While many homes were also in a general state of disrepair, many men continued to harvest poles to rebuild their homes regardless of the formal institutions that prohibited the felling of trees for poles in the same year. The most pronounced defiance by all forest residents in Gwayi West is their continued stay despite the official policy and practices to relocate them to the east since 1993. Even with the power relations embedded in threats and practices of "harassment" by successive foresters, forest residents continue to live in the area they are supposed to leave.

The way different actors talked about various FC officials represented another form of resistance. Language in public arenas rarely expresses hidden agendas (Scott 1990), except in some circumstances in Gwayi West when residents openly denied being called "squatters." Away from forest officials, however, residents expressed decidedly different opinions about prevailing policy and practice. For example, one old man whose family had been moved from the Insuza Wetland to make way for a wildlife habitat in the 1960s said: "Mrs. Farqhuar [the former safari camp manager in Gwayi Forest], who came and took land away from us, can kill animals at will and cut trees for pleasure, while we cannot hunt to feed ourselves, and yet she found us living here. This is very painful!"[9]

Covert resistance also took the form of pilfering and theft (as with the second type of resistance) of prohibited resources from the forest. Sometimes different households do the pilfering to "fix" forest authorities and assert a group's rights to forest resources or just as a means to gain access to highly valued resources. Recourse to such acts of defiance illustrates the agency that local people exercise in the face of neoliberal conservation practices. This agency is exercised in a variety of complex ways linked to both temporal and spatial contexts that the cases presented here illustrate, even though there are structurally dominant ways in which the state exercised power in the forests, particularly the Foucaultian senses of discipline (in relation to ordering people to live their lives in certain ways that were not normal or natural to them), discourse (not only were people termed "squatters" when they had lived in forests before the FC, but local people's knowledge was considered irrelevant in managing forests), and governmentality (through Scott's ordering of nature in certain ways and, together with these practices, the way people lived their lives to conform to this view of how forests should be). The cases presented here illustrate the feedback that local people provided to the state's power (Giddens's structuration [1984]; Scott 1985, 1990) by exercising their own power and agency.

7. Conclusions

This chapter has presented the argument that in Zimbabwe, the state, through the Forestry Commission, has been setting aside forest landscapes for purposes of conservation—ostensibly a public good, but in reality this public good has been used to further the state's strategy of accumulation by dispossession (Harvey 2006b). This state accumulation has been done at the expense of local people who have been relegated to even poorer forms of existence to make way for the state to collude with private capital through the various conservation practices elaborated throughout the chapter. The chapter has also presented the outcomes of these processes of primitive accumulation, outcomes that lead to local people engaging in various forms of resistance.

In a Foucaultian sense, such resistance constitutes an exercise of power that, in most circumstances, takes more "quotidian" forms (Scott 1985). That is, the exercise of power is more observable in the nature of relations between forest authorities and forest residents and neighbors in the day-to-day struggles over control of forest resources and the assertion of identity, respectively. As described earlier in the chapter, policy framing represents a way of exercising power in Foucaultian terms (Foucault 1982). Through the agency of local actors, these conservation practices are resisted, leading to contentious relations among the different stakeholders around forest resources. Forest reserves are thus symbolic arenas for realizing divergent visions and livelihood patterns. Early inhabitants of the forests clash with forest authorities to assert their rights of stay (for forest residents), as their ways of life are intimately connected to various forest resources. Access to or use of some of these resources is nevertheless prohibited by the FC because of the value of these resources to the bureaucracy, especially under neoliberal dictates, where the central state has withdrawn support for conservation activities. Thus we have seen, for example, how the state protection of Gwayi Forest as a wildlife habitat has marginalized local people's livelihood needs and sense of identity concerning forest use. This has resulted in the development of resistance by forest residents as they exercise their power and agency.

This chapter has presented the complex interaction between neoliberal conservation practices by the state and the resistance of forest-dependent people. The chapter has therefore illustrated the different forms of agency exercised by local people in response to the privatization of the state in forestry conservation. Such different types of agency are not commonly recognized in either the literature on neoliberal conservation or that addressing accumulation by dispossession, given their emphasis on broader issues of political economy. The findings have also made a contribution toward understanding the effects of the rolling back of the state raised by Ferguson (2006) and the impact of this process on different local people in various settings who do not just sit back in the face of this exercise of power but contest, resist, and struggle against it in many ways.

Notes

1. Letter by the area manager to the manager, Indigenous Resources Division of the Forestry Commission of Zimbabwe, "Minutes of the Tenants Meeting with Minister, J. Msika, on the 2nd of February, 1991."

2. Interview with Gwayi West chairman of the Residents Association after the beating incident on September 16, 1997.

3. Expression from interview with Gwaqula in Gwayi West, September 1998.

4. One resident's reaction to hard treatment by foresters in Gwayi West, November 1997.

5. The chairman of the Residents Association as a local ruling party representative could access the governor's office at any time through party channels.

6. Interview with Matebeleland North governor in Bulawayo over his opinion about the conflicts in Gwayi, October 1998.

7. Interview with Heneri, August 1998.

8. Interview with Mr. Ndhlovu in Gwayi East, September 1998.

9. An elderly resident's statement given in the middle of an interview in Gwayi West, September 1998.

PART II

Representations of Nature™ Inc.

CHAPTER FOUR

Taking the Chocolate Laxative

Why Neoliberal Conservation "Fails Forward"

Robert Fletcher

As the full scale of the ecological problems confronting us has become increasingly apparent, efforts to address them have become increasingly focused on engagement with capitalist markets—a trend described in this volume as "Nature™ Inc." Yet for many critics, it is precisely capitalist markets that are in no small part responsible for the environmental problems they are now called upon to solve (see Fletcher 2012c). Büscher (2012, 29) thus describes neoliberal conservation as "the paradoxical idea that capitalist markets are the answer to their own ecological contradictions." Consequently, while the ecological crisis threatening the future of life on Earth is becoming increasingly acknowledged within mainstream global society, at the same time it is becoming increasingly difficult to imagine addressing this crisis in other than capitalist terms (see also Swyngedouw 2011), reinforcing Jameson's (2003, 76) observation that "it is easier to imagine the end of the world than it is to imagine the end of capitalism." Accounting for this paradox is the purpose of this analysis.

It does so by addressing two interrelated questions. First, why is there so often a substantial gap between theory and practice in neoliberal forms of environmental conservation? As I will show, many ostensibly market-based mechanisms include in their actual implementation tools and strategies largely antithetical to foundational neoliberal principles. In addition, irrespective of their methods, such mechanisms often fail to achieve intended results. While these dynamics have been noted by other research with respect to specific contexts and strategies, I suggest that they actually constitute a common pattern in neoliberal conservation. I contend that this situation results in part from the particular virtualistic (Carrier and Miller 1998; Carrier and West 2009) vision underlying market mechanisms, insofar as they seek to transform the world to conform to a model that is assumed to already exist. This vision, I suggest, contains fundamental errors while offering such impossible criteria for fulfillment that reconciling theory and practice would in fact be quite difficult.

If this is so, however, it raises a second key question: Why is this gap between vision and execution so rarely acknowledged within neoliberal discourse? I demonstrate that, despite a common failure to execute neoliberal conservation strategies as envisioned, this reality is seldom directly attributed to the fundamental nature of market mechanisms themselves. Rather, it seems, neoliberal analysts tend to engage in a sort of "fetishistic disavowal" (Žižek 1989, 2008)—a simultaneous admission and denial—often superficially acknowledging yet ultimately dismissing for the most part potential critiques concerning the presence of essential contradictions in the operation of neoliberal mechanisms.

I contend that this dynamic is sustained through recourse to fantasy, by means of which the gap between what Carrier and West (2009) call "vision" and "execution" (or, in Lacanian terms, between Real and Symbolic) is effectively obfuscated. As with all fantasies, neoliberalism is supported through stimulation of desire, in this case, desire for fulfillment (both material and sensual) via the spectacular consumption that neoliberal capitalism characteristically promotes. Such fulfillment, however, is most commonly denied, replaced by what Neves (2009a) calls "pseudocatharsis," a partial satisfaction that paradoxically augments the very desire it purports to satiate by providing subjects with just enough stimulation that they are motivated to continue their commitment to neoliberal mechanisms in quest for the resolution constantly deferred. In this way, the failure of neoliberal policies to achieve intended aims is disavowed and faith in the neoliberal project sustained.

In developing this analysis, I build on a growing body of research addressing neoliberalization within natural resource management generally (see, e.g., Castree 2008a, 2008b, 2010a, 2010b; Heynen et al. 2007) and conservation specifically (e.g., Brockington, Duffy, and Igoe 2008; Brockington and Duffy 2010a, 2010b; Büscher et al. 2012; Corson, MacDonald, and Neimark 2013; Fairhead, Leach, and Scoones 2012b; Fletcher 2010, 2012c; Sullivan 2006a; but see Büscher et al. 2012 for insightful discussion of important similarities and differences between these interrelated literatures). At the same time, I draw on Žižek's (e.g., 1989, 2008) idiosyncratic fusion of Marx and Lacanian psychoanalysis to take this discussion in new directions. While Žižek has employed his framework to comment extensively on capitalist mechanisms in general, the particular dimensions of neoliberalization have not been extensively addressed. Several recent works have drawn on Žižek and other psychoanalytic perspectives to engage with neoliberalism in general (e.g., Dean 2008; Layton 2009; Glynos 2012) but have not addressed environmental governance. Likewise, several researchers have analyzed environmentalism from a Lacanian/Žižekian perspective (e.g., Stavrakakis 1997a, 1997b; Swyngedouw 2010, 2011; Kingsbury 2010, 2011) but have left the particular dimensions of neoliberalization within environmental governance largely unexplored. By bringing all of these discussions together (in a theoretical framework described most extensively in Fletcher 2013), I aim to produce a novel synthesis that contributes to the understanding of the growing trend to employ free-market mechanisms to mitigate environmental degradation. As Bloch observes, "Most critics of neoliberalism leave the reader mystified as to how such flawed ideas could ever have become so powerful" (in Peck 2010, back cover). Building on Peck's (2010) own incisive analysis, the present discussion seeks to account for this dynamic with respect to neoliberal conservation specifically.

I begin by outlining the Lacanian/Žižekian framework underpinning this analysis. I then apply this framework to highlight a common divergence between vision and execution in the implementation of neoliberal environmental governance mechanisms, observing a characteristic pattern of antineoliberal intervention in the face of market mechanisms' common failure to perform as envisioned. I document a widespread tendency to deny this reality by explaining it not as a failure of market mechanisms per se but, on the contrary, as a failure to engage the market sufficiently. I analyze this as an expression of disavowal concerning the contradictions of neoliberal governance, describing the manner in which the logic of neoliberal discourse obfuscates these dynamics. Finally, I apply this framework to an analysis of what is arguably the most prominent form of neoliberal conservation globally at present, namely, ecotourism.

Neoliberalism, Fantasy, and Desire

Žižek's framework (first substantially outlined in *The Sublime Object of Ideology* [1989]) is grounded in Lacan's iconic triad: Imaginary–Symbolic–Real. In this model, the *Real* is a placeholder name for that which subverts signification, exhibiting a dual character as "both the hard, impenetrable kernel resisting symbolization and a pure chimerical entity which has in itself no ontological consistency" (Žižek 1989, 190). By contrast, the *Symbolic* designates our attempts to represent the Real and impose order upon it. Due to the very nature of the Real, however, such representation inevitably falls short of its aim. The Real, as Lacan famously asserted, is thus "impossible," incapable of representation; it is "the rock upon which every attempt at symbolization stumbles" (190).

As a result, there is invariably a gap between the Real and its Symbolic representation, with the Real comprising an "irreducible excess" overflowing our illusions of order and coherence. This excess, denied within the symbolic order, manifests as "symptom," the "return of the repressed" (Žižek 1989, 57) by means of which the Real ruptures and undermines Symbolic attempts to create coherence. A symptom is thus "the point at which the immanent social antagonism assumes a positive form, erupts on to the social surface, the point at which it becomes obvious that society 'doesn't work,' that the social mechanism 'creaks'" (143). A symptom therefore indicates a fundamental antagonism or inconsistency in the social order; it is a "surplus-object" or "the leftover of the Real eluding symbolization" (51).

The *Imaginary*, the third element in Lacan's triad, represents our efforts to conceal this essential disjuncture by means of fantasy, which Žižek calls the "screen concealing the gap" (1989, 132) between Real and Symbolic. Fantasy thus "constitutes the frame through which we experience the world as consistent and meaningful," obscuring the fact that the Symbolic order is in fact "structured around some traumatic impossibility, around something which cannot be symbolized" (138). In other words, "fantasy is a means for an ideology to take its own failure into account in advance" (142).

This function of fantasy is sustained through desire, pursuit of what Lacan called *jouissance*, usually translated as "enjoyment" but more properly a mixture of pleasure and pain that promises a satisfaction it can never deliver. Hence, unresolved desire is sustained over time, and thus "in the fantasy-scene desire is not fulfilled, 'satisfied,' but constituted" (Žižek 1989, 132). Rather, "through fantasy, *jouissance* is domesticated" (138). In this way, fantasy's promise to deliver the desired satisfaction at some future point conceals the impossibility of this promise, the Real–Symbolic gap it obscures, and the symptoms that signal this disjuncture as well. In the process, fantasy commonly invokes a scapegoat, such that the disjuncture between Real and Symbolic is further sutured by positing the infiltration of "an external element, a foreign body introducing corruption into the sound social fabric" (142). Consequently, "what is excluded from the

Symbolic . . . returns in the real as a paranoid construction of" that which is repressed (143).

As a shorthand description for this complex dynamic, *ideology* can thus be understood as "a totality set on effacing the traces of its own impossibility" (Žižek 1989, 50).

In applying his framework to social analysis, Žižek outlines two "complementary procedures" comprising a comprehensive psychoanalytic inquiry: "One is *discursive*, the 'symptomal reading' of the ideological text bringing about the 'deconstruction' of the spontaneous experience of its meaning. . . . [T]he other aims at extracting the kernel of *enjoyment*, at articulating the way in which—beyond the field of meaning but at the same time internal to it—an ideology implies, manipulates, produces a pre-ideological enjoyment structured in fantasy" (1989, 125, emphasis in original). In the following, I apply these twin procedures to analyze the peculiar persistence of neoliberal conservation (Fletcher 2013) in both discursive and visceral dimensions.

How Neoliberal Is Neoliberal Conservation?

As previously described, so-called market-based conservation mechanisms are increasingly promoted to address pressing environmental problems, from biodiversity loss to climate change, by international financial institutions (IFIs), national governments, private-sector firms, and, increasingly, even the NGOs ostensibly intended to represent civil society interests vis-à-vis all of the preceding (Levine 2002; Chapin 2004; Corson 2010). Such market mechanisms, which include established forms like ecotourism, payment for environmental services (PES), and carbon markets (Fletcher 2009, 2012b; Fletcher and Breitling 2012), as well as newly emerging forms including environmental derivatives and species and wetlands banking (see Sullivan 2013b; Büscher, this volume), are explicitly designed to ascribe sufficient monetary value to natural resources that stakeholders will elect to preserve rather than deplete them, thereby incentivizing conservation over resource extraction (see Fletcher 2010).

Despite this growing enthusiasm concerning neoliberal conservation mechanisms, their efficacy in many cases remains questionable. Indeed, a growing body of research demonstrates that such mechanisms in fact commonly fail to perform as intended (e.g., Büscher and Dressler 2007; Carrier and West 2009; Fletcher 2012c; Fletcher and Breitling 2012; McAfee 2012a, 2012b; Lohmann 2011; Milne and Adams 2012; West 2006). Moreover, in cases where neoliberal mechanisms do appear successful, there is often ambiguity concerning the extent to which such mechanisms are actually faithful to the principles they claim to enact. Even acknowledging the extreme heterogeneity of neoliberalization in practice (Büscher and Dressler 2012; Dressler and Roth 2010; Harvey 2005), an essential neoliberal tenet dictates that primary responsibility for allocating resources should be left to market actors, with the state acting mostly to provide the legal and administrative structures shaping markets rather than intervening directly (see Foucault 2008; Fletcher 2010). This is certainly not to imply that neoliberalism proscribes all state involvement in the market, as critics often assume. On the contrary, Foucault (2008, 146) observes, foundational neoliberal economists, including both Hayek and Friedman, maintained that intervention was in fact necessary to "make the market possible," for the market was not seen as a natural, presocial entity but as an artificial construct requiring continual maintenance (see also Fletcher 2010; Peck 2010). Hence, Foucault (2008, 132) maintains, "Neoliberalism should not be identified with laissez-faire, but rather with permanent vigilance, activity and intervention."

Even given this more nuanced understanding of the neoliberal state, however, it is clear that many ostensibly neoliberal conservation mechanisms do not function as such in practice. In his analysis, Foucault (2008, 174) distinguishes between "organizing" and "regulatory" actions, observing that while neoliberalism encourages the former (through state action to create the "conditions of the market via legal and administrative structures"), it seeks to minimize the latter actions, which entail direct intervention in market transactions. Hence, mechanisms that involve substantial regulatory intervention rather than mere organizing activity can be seen as contrary to core neoliberal principles.

My previous research has revealed several instances in which ostensibly neoliberal conservation strategies function otherwise in practice. Here, I will consider just two of these, ecotourism and PES, as these are arguably the most paradigmatic (and widespread) neoliberal conservation mechanisms at present. Concerning PES, research demonstrates that Costa Rica's celebrated national program, while widely hailed as a successful example of market-based conservation, actually relies extensively on nonmarket mechanisms to function, including a national law prohibiting land-use change on private land and national fuel and water tariffs appropriating and redistributing resources to fund the system (see Fletcher and Breitling 2012). In fact, voluntary market transactions, upon which the program is ostensibly based in its expressed intention to transfer payments from consumers of environmental services to these services' producers, comprise less than 1 percent of the program's total activity, despite persistent efforts to enhance the centrality of this mechanism (Blackman and Woodward 2010). As a result, over the program's lifetime the state has been forced to increase its reliance on nonmarket tools in order to keep the program running, with the result that, in its actual practice, the PES system functions less as a neoliberal mechanism than as a subsidy in disguise (Fletcher and Breitling 2012). Milne and Adams (2012) reach a remarkably similar conclusion concerning a community-based PES program in Cambodia.

My research on ecotourism demonstrates an analogous dynamic. While industry advocates often explicitly endorse a neoliberal approach to encouraging local participation in the industry, promoting ecotourism as a means for local users to profit from in situ natural resources and thus incentivizing conservation (what Honey [2008, 3] calls the "stakeholder theory"), in their actual practice planners tend to enact a quite different perspective. Planners' efforts to appeal to locals commonly entail an implicit effort to acculturate the latter to aspects of the particular cultural perspective that motivates ecotourism's practice, demonstrating that despite their neoliberal rhetoric, many planners actually consider such value change as important to successful ecotourism development as provision of economic incentives (Fletcher 2009).

Similar dynamics have been documented elsewhere (e.g., Carrier and West 2009; Büscher and Dressler 2007). Hence, Büscher and Dressler (2007) describe a common gap between "reality" and "rhetoric" in neoliberal environmental policies, while Carrier and West (2009) identify a similar disjuncture between "vision" and "execution." On the whole, Harvey (2005, 19) observes "a creative tension between the power of neoliberal ideas and the actual practices of neoliberalization that have transformed how global capitalism has been working over the last three decades." Peck (2010, xiii, emphasis in original) describes the history of neoliberalization as "one of repeated, prosaic, and often botched efforts to *fix* markets, to build quasi-markets, and to repair market failure." Steger and Roy (2010) outline a long series of failed efforts of neoliberal policies to achieve intended results in diverse contexts, from Chile's dramatic 1982 recession following nearly a decade of aggressive liberalization through the second US President Bush's plunging of the global economy into the current crisis. Indeed, one might assert with Peck (2010) that a neoliberal vision-execution gap is nigh inevitable, that neoliberal conservation mechanisms may in fact be largely incapable of achieving their lofty goal of facilitating substantial resource preservation on a global scale. Why this would be so is the subject of the next section.

Vision and Execution

In his influential *Brief History of Neoliberalism*, Harvey contends that neoliberal economics functions largely as an ideological smokescreen concealing the fact that neoliberalization is at root a project by means of which a transnational capitalist class (Sklair 2001) seeks to consolidate wealth and power through "accumulation by dispossession": "It has been part of the genius of neoliberal theory to provide a benevolent mask full of wonderful-sounding words like freedom, liberty, choice, and rights, to hide the grim realities of the restoration or reconstitution of naked class power, locally as well as transnationally, but most particularly in the main financial centres of global capitalism" (Harvey 2005, 119).

Neoliberalism here is thus understood as an ideology in the classic Marxist sense of a "false consciousness" concealing an underlying objective reality (see Scott 1990; Fletcher 2007). From this perspective, the ostensive "gap" between vision and execution in neoliberal governance exists because neoliberal policies were never in fact intended to function in the public interest in the first place. Hence, Harvey (2005, 19) asserts that "when neoliberal principles clash with the need to restore or sustain elite power, then the principles are either abandoned or become so twisted as to be unrecognizable."

An alternate yet complementary view takes its cue from Polanyi's (1944) analysis of liberalism's "double movement" to suggest that neoliberal policies tend to produce such "perverse economic consequences and pronounced social externalities" that extramarket intervention is required to redress these excesses in order to stave off the social unrest they would otherwise provoke (Peck and Tickell 2002, 388). In this frame, the vision-execution gap results not merely from the project of accumulation per se but from the need to restore order when this project runs awry (see also Peck 2010).

Undoubtedly there is some truth in both of these analyses, instances in which elites cynically manipulate neoliberal policies for their own ends while subordinates embrace the dominant ideology offered to conceal this aim, as well as moments in which neoliberal excess demands a reactionary response in the form of state intervention. However, as might be expected given his trenchant critique of many aspects of Marxist epistemology (see esp. Foucault 1991), in a recently published series of lectures from 1979 Foucault (2008, 218) preemptively contests the perspectives outlined above, viewing neoliberalism not as an ideological smokescreen concealing material interests but as a discourse, that is, as a "whole way of thinking and being," a "general style of thought, analysis and imagination." In this understanding, neoliberalism does not conceal so much as construct reality—or at least a certain depiction of reality—by generating a particular "truth-regime of the market" (144). This in fact brings us close to Žižek's understanding of ideology: it is "not simply a 'false consciousness,' an illusory representation of reality, it is rather this reality itself which is already to be conceived as 'ideological'" (1989, 15).

Neoliberalism, in this sense, can be understood as a quintessential example of what Carrier and colleagues (Carrier and Miller 1998; Carrier and West 2009) call "virtualism," a project that seeks to reshape the world in conformance with its predetermined vision while claiming to merely reflect the reality it seeks to transform. Hence, Lemke (2001, 203) describes neoliberalism as a "political project that endeavours to create a social reality that it suggests already exists," while Bourdieu (1998b) calls neoliberalism "the implementation of a utopia" that paradoxically "conceiv[es] of itself as the scientific description of reality."

From this perspective, then, neoliberalism would not be conceived as a mystifying ideology obscuring the "true" aim of class consolidation but would take at face value that at least some neoliberal advocates earnestly believe that their perspective is capable of achieving intended goals and genuinely strive to make it do so (Li 2007; Peck 2010). This is certainly not to deny that the philosophy can be used cynically at times to pursue actors' self-interested ends (as neoliberal theory itself would indeed predict) or conceal their true intentions, merely that one cannot necessarily assume such motives at the outset in all cases. In this understanding, neoliberal policies would be seen to fail to achieve intended aims when their virtualistic vision conflicts with the reality it seeks to transform. Büscher and Dressler (2007) therefore attribute disjuncture between rhetoric and reality in neoliberal conservation to the fact that projects tend to operate with simplistic blueprints that do not do justice to the complex local realities they confront, echoing Scott's (1998) influential analysis of the common failure of large-scale transformational projects in general. In a Lacanian frame, this would be described as a manifestation of the inevitable gap between Real and Symbolic representation.

I would go further, however, to contend that neoliberal projects fail not merely due to their simplified vision per se but due to the particular nature of this vision. First, there is obviously the fact that, as a number of critical economists recognize, the so-called free market that neoliberal policies seek to implement and harness is extremely difficult—if not impossible—to realize. As Stiglitz (2008b, 42) describes, "Markets were efficient only if capital markets were impossibly perfect. . . . There could be no externalities (no problems of air and water pollution), no public goods, no issues of learning, and no advances in technology that were the result either of learning or expenditures on R&D. . . . [T]here also could not be any imperfections of information, changes in the information structure, or asymmetries of information."

In addition to these significant obstacles, the theory of essential human nature—and thus of the nature of motivation, human relations, and social change as well—in which neoliberalism is grounded is, I would argue, so fundamentally flawed, misrepresenting in such significant ways how humans actually behave, that it would in fact be quite difficult for neoliberal policies to function as intended. Specifically, neoliberal theory is commonly grounded in a rational actor/*Homo economicus* model of human nature and behavior, assuming that people seek, first and foremost, to maximize their material utility and perform cost-benefit assessments in order to do so (see Foucault 2008). The basic neoliberal governance strategy—to create incentive structures that encourage people to exercise their rational choice in socially desired ways—follows fundamentally from this view (Fletcher 2010). Yet this model of human nature has long been contested by anthropologists, among others, who contend that it does not accurately describe human motivation in all situations and contexts (see Büscher et al. 2012 for an overview of this critique). Graeber (2011), indeed, contends that only through centuries of concerted violence, both physical and conceptual, has this peculiar vision become so widely accepted. And if the understanding of the nature of human motivation informing neoliberal policies is so essentially flawed—if there is such a fundamental gap, in other words, between the Real of human nature and its representation in neoliberal theory—policies and practices following from this understanding would be hard-pressed to function otherwise.

Disaster and Dissimulation

Notwithstanding the frequency of its occurrence, neoliberalism's failure to achieve intended results is rarely explicitly acknowledged—is, in fact, commonly denied or explained away—within mainstream economic discourse (Peck 2010; Büscher et al. 2012). Consider, for instance, the dominant response to the current global economic crisis. The deepening recession of late 2008 witnessed a proliferation of statements claiming that the crisis signaled neoliberalism's wholesale failure and the need to return to some type of Keynesianesque regulatory regime (see Peters 2008). Stiglitz

(2008a), most trenchantly perhaps, proclaimed: "Neo-liberal market fundamentalism was always a political doctrine serving certain interests. It was never supported by economic theory. Nor, it should now be clear, is it supported by historical experience. Learning this lesson may be the silver lining in the cloud now hanging over the global economy."

Yet less than a year later, following the infamous bailout packages implemented in a number of countries, such voices had largely fallen silent or even reversed their positions. Thus the IMF (2009, 9, emphasis added) stated in its annual report: "The seeds of the global crisis were sown during the years of high growth and low interest rates that bred excessive optimism and risk taking and spawned a broad range of failures—in *market discipline, financial regulation, macroeconomic policies, and global oversight*." The last three of these "failures," of course, refer to errors in state governance, while the first concerns the "irrational exuberance" of market players (a classic neoliberal scapegoat—see Dean 2008; Layton 2009). Similarly, an open letter signed by more than two hundred economists published in the *New York Times* in January 2009 asserted, "Lower tax rates and a reduction in the burden of government are the best ways of using fiscal policy to boost growth" (cited in Peck 2010, 270).

All of this conforms to Foucault's (2008, 116) characterization of the logic of neoliberal response to economic crisis in general as maintaining, "Nothing proves that the market economy is intrinsically defective since everything attributed to it as a defect and as the effect of its defectiveness should really be attributed to the state." A similar response had predominated in past neoliberal crises as well. Harvey (2005, 97) describes this same logic in the mainstream response to the 1997 Asian financial collapse, observing, "The standard IMF/US Treasury explanation for the crisis was too much state intervention and corrupt relationships between state and business ('crony capitalism'). Further neoliberalization was the answer." Steger and Roy (2010) identify an analogous reaction to the series of crises precipitated by neoliberal restructuring in the United States and elsewhere over the span of several decades. In short, Peck (2010, 6) observes, "It is both an indictment of neoliberalism and testament to its dogged dynamism, of course, that laboratory experiments do not 'work.' They have nonetheless tended to 'fail forward,' in that their repeated manifest inadequacies have—so far anyway—repeatedly animated further rounds of neoliberal intervention."

This response, which I call "the beatings will continue until morale improves" strategy, has been prevalent within the realm of environmental governance as well. In October 2008, as the crisis deepened, the International Union for Conservation of Nature held its World Conservation

Congress in Barcelona, Spain, during which, as Ken MacDonald (2010a) describes, a market agenda that had until then been hotly debated among the organization's membership achieved hegemony within the upper echelons. Christine MacDonald (2009) thus observes: "By early October, when the world's conservation elites gathered in Barcelona for their biggest meeting of the year, markets were crashing around the world, spreading panic and doubt about the wisdom of unbridled free market economics. But the conservationists, corporate CEOs, billionaire philanthropists, and heads of state and royal houses don't seem to have heard the news. In Barcelona's conference rooms and banquet halls, the conversation centered on how environmental groups must become even more like corporations."

How do we account for this persistent resilience of neoliberal ideology in the face of widespread critique concerning its failure in practice? In Harvey's Marxist vision, of course, this is easily explained: if neoliberal ideology is not intended to accurately account for reality at all but merely to obfuscate the class project it seeks to legitimate, then "failure" is merely further obfuscated through additional obscuring rhetoric. Undoubtedly, again, there is some truth in this view, at least with respect to some (perhaps most) stakeholders. Yet this appears to be only part of the picture. Büscher and colleagues (2012, 24) offer a more nuanced analysis of this issue in seeking to explain "why the neoliberalizing of environmental conservation is so opaque and seductive to those involved with conservation work." They contend that neoliberal conservation tends to obfuscate fundamental contradictions in its deployment by means of three main tactics: (1) "win-win" rhetoric asserting that these contradictions can be resolved through the very processes that stimulate them; (2) media spectacle creating "the *appearance* of general consensus with the ideological assumptions of neoliberal capitalism" (Büscher et al. 2012, 18, emphasis in original); and (3) the aggressive disciplining of dissenting voices that question these representations. All of this, they contend, results in a sort of "closed loop" thinking, "whereby in failing to take into account the wider processes of which it is part, the self-corrective actions of an ill-functioning system perpetuate illness-causing conditions, while providing temporary illusion of improvement" (14).

Yet, again, this may explain only a portion of the phenomenon in question. Consider the following statement, offered by an ecological economist in the online discussion of a critique of emerging conservation financialization strategies (Sullivan 2010b): "If we were serious about having a true market economy, mergers and acquisitions and other means of concentrating power would be disallowed" (William Rees, http://www.capitalinstitute.org/forum/ecosystem/can-nature-be-monetized-capital-institute-conversation, accessed October 15, 2010). This statement strikes me as extraordinary, tantamount to acknowledging an inherent contradiction at the very heart of neoliberal policy. For free markets to exist, Rees appears to be saying, they must be regulated in ways diametrically opposed to the essential neoliberal principle dictating that the state should not intervene directly in the market to regulate players' resource allocation decisions. This contradictory pronouncement seems to go beyond the techniques of obfuscation that Büscher and colleagues (2012) highlight. Rather, it smacks of the dynamic that Žižek (1989, 12) calls "fetishistic disavowal," represented by the formula: "I know very well, but still . . ." In Žižek's (2008, 14) description, for instance, a Cold War Communist might admit, "I know very well that things are horrible in the Soviet Union, but I nevertheless believe in Soviet socialism." Similarly, in the statement cited above, Rees seems to be saying, "I know free markets are impossible; nevertheless, I believe that they can function."

Likewise, Büscher and colleagues (2012, 15) note that "not all conservationists are so smitten by the allure of neoliberal solutions," highlighting instances in which critical assessments of market-based mechanisms have been published by prominent mainstream voices in core conservation journals (e.g., Chan et al. 2007; Child 2009; Ehrenfeld 2008; Peterson et al. 2009; Redford and Adams 2009; Walker et al. 2009). Yet such "critical messages are often ignored by mainstream organizations and media, and if they are acknowledged, often denied or twisted to suit particular neoliberal objectives" (Büscher et al. 2012, 22). As a result, the authors observe, paradoxically, that "alternative viewpoints do not always need to be actively suppressed in order to be disciplined. Indeed, they can perversely be stimulated as some kind of catharsis, without impacting on the broader hegemonic system" (22).

A Lacanian/Žižekian analysis may help to shed light on this dynamic by framing neoliberal theory and the practices through which it is implemented as a Symbolic order that attempts to simultaneously represent and shape the Real in virtualistic fashion. The Real, of course, inevitably exceeds this imposition, generating the characteristic gap between vision and execution in neoliberal governance observed earlier. This gap manifests as symptom, the main form of which, as Žižek (2008) describes of capitalism in general, is superfluous *waste*: the environmental and social excess that neoliberalism externalizes in its quest for profit, namely, the ecological damage wrought by and the masses rendered expendable within the capitalist production process.

The gap between neoliberal theory and the Real it confronts, as well as the waste accumulated as symptom of this gap, is sutured within the Imaginary through fantasy. Central to neoliberal environmental governance, from this perspective, is the fantasy that "capitalist markets are the answer to their own ecological contradictions" (Büscher 2012, 29; see also Swyngedouw 2010, 2011). This, of course, is the underlying logic of all of the governance mechanisms described at the outset, functioning like Žižek's oft-quoted example of the chocolate laxative, a real product that claims to be the antidote to the problem (constipation) it itself provokes. This conviction is grounded in two further fantasies: that by addressing environmental and social issues in the production and consumption process, capitalism can facilitate (1) accumulation without end and (2) consumption without negative consequences. In this second fantasy, so-called ethical consumption claims to resolve the contradiction between the increased consumption essential to capitalist expansion and the ecological/social crisis provoked by this expansion by ostensibly linking purchase to social programs that claim to actually redress rather than stimulate such crisis (Carrier 2010; Igoe 2010; Igoe, Neves, and Brockington 2010; West 2010). In the first fantasy, meanwhile, market environmentalism claims to resolve the parallel opposition between economic growth and environmental limits by promoting ostensibly sustainable—even "nonconsumptive" (West 2006)—resource exploitation (Brockington, Duffy, and Igoe 2008; Büscher et al. 2012).

The gaps between Real and Symbolic in all of these constructions have been increasingly highlighted by critical analysis. "Ethical consumption" has been shown, in many cases, to carry substantial social and environmental consequences that are obscured by self-congratulatory rhetoric claiming universally beneficial outcomes (Carrier 2010; Igoe 2010; Igoe, Neves, and Brockington 2010; West 2010). The common recourse to celebrity endorsement for environmental causes and products aids such obfuscation (Brockington 2009; Sullivan 2011b). Market environmentalism, meanwhile, entails its own contradictions, leading conservationists to increasingly conclude that its aim to reconcile economic development and environmental protection has largely failed thus far (see, e.g., McShane et al. 2011)—without, however, attributing this failure to market mechanisms themselves but rather to ostensibly inherent "trade-offs" between conservation and development concerns (see Fletcher 2012c).

All of this, however, is further mystified by means of the metafantasy at the heart of neoliberal governance in general, which Dean (2008, 55), drawing on Žižek, calls the "fantasy of free trade": "The fantasy of free trade covers over persistent market failure, structural inequalities, the violence of privatization, and the redistribution of wealth to the 'have mores.' Free trade sustains at the level of fantasy what it seeks to avoid at the level of reality—namely, actually free trade among equal players, that is equal participants with equal opportunities to establish the rules of the game, access information, distribution, and financial networks, etc." By means of this fantasy, any shortcomings of neoliberal policies can be explained away via the logic that these problems are due not to neoliberalism itself but to the failure to implement it properly and to a sufficient degree. Hence Foucault's (2008, 116) characterization of neoliberal reasoning, cited above, as asserting that "nothing proves that the market economy is intrinsically defective since everything attributed to it as a defect and as the effect of its defectiveness should really be attributed to the state." In other words, failure is not failure of market logic but merely "market failure," a technocratic rather than political problem (Li 2007; Büscher 2010a) remedied by simply working harder to "get the market right." Hence, the fantasy of free trade dictates that no matter how greatly neoliberal policies fail in practice, the fiction can be sustained that this failure is due not to any fundamental errors or contradictions in these policies' internal logic but rather to the fact that they have been implemented incompletely; thus, if only neoliberalism could function with less inhibition, less state regulation, it would actually perform as intended. As Žižek observes, in this fantasy (neoliberal) capitalism fails because it is not pure *enough*, reinforcing his observation that "ideology really succeeds when even the facts which at first sight contradict it start to function as arguments in its favour" (1989, 50). In this way, critique of neoliberal logic itself is effectively neutralized.

The *Objet Petit A* of Desire

As Dean (2008) points out, it is of course desire that sustains the fantasy of free trade: desire on the part of neoliberal advocates to see their policies function as intended; desire on the part of those excluded from neoliberalism's benefits to finally receive the material rewards dangled in front of them. De Vries (2007) identifies this latter function of desire in international development policy, wherein the masses excluded from the fruits of development may nevertheless sustain faith in development's potential due to their desire to receive the benefits (projects, public works, etc.) long promised by planners (see also Glynos 2012). It is this same desire, I suggest, that in part makes neoliberalism so resilient, so resistant to critique; rather than undermining the perspective, paradoxically, neoliberalism's thwarting of the *jouissance* it promises merely enhances its appeal by augmenting desire for the elusive fulfillment. In addition, of course, neoliberalism does in fact deliver a semblance of a promised pleasure, and thus "for quite a few people, capitalism is not just hard graft. It is also fun. People get stuff from it—and not just more commodities. Capitalism has a kind of crazy vitality. It doesn't just line its pockets. It also appeals to gut feelings" (Thrift 2005, 1; see also Glynos 2012). Yet this may merely enhance one's ideological attachment by stimulating desire for further sensation in pursuit of a satisfaction constantly deferred (Fletcher and Neves 2012).

To further illuminate this process it may help to return to Žižek and Lacan, who claim that the desire stimulated by fantasy originates in a fundamental "lack" intrinsic to subject formation for which the pursuit of *jouissance* seeks in a sense to compensate. Freud (1962) described an "oceanic feeling," which he claimed represented a reminder, in a sense, of the primordial sense of oneness one experiences during preseparation infancy. Echoing this, Lacan distinguished *jouissance* (lowercased) from *Jouissance* (capitalized), describing the latter as akin to Freud's oceanic feeling: a primordial sense of wholeness that subjects attempt to recapture later in life in the form of a lesser, derivative *jouissance* (see Fink 1995). As a result of its origin in an essential lack, however, *jouissance*, as noted earlier, promises a satisfaction it can never deliver, ensuring, paradoxically, that unresolved desire is sustained rather than resolved, for as Lacan asserted, desire is always at root a desire for desire—what Lacan termed the impossible *objet petit a*—itself. Indeed, Žižek contends that the very idea of a primal *Jouissance*, which present *jouissance* seeks to replicate, is in fact illusory, a "fantasmatic construction" sustaining the delusion that "there was once a time or space before lack" (McMillan 2008, 22).

All of this coalesces in the peculiar configuration of neoliberal environmental governance, in which contradictions are concealed and faith is sustained through fantasies of future fulfillment of the desire paradoxically stimulated by neoliberal arts of government themselves. This is well illustrated by the example of ecotourism, probably the most widespread form of neoliberal environmental governance in the world at present (see, e.g., Duffy 2002, 2008, 2012; Duffy and Moore 2010; Fletcher 2009, 2011; Fletcher and Neves 2012). As described in detail elsewhere (Fletcher 2011; Fletcher and Neves 2012), the tremendous attention devoted to the potential of ecotourism to solve myriad environmental and social ills over the past several decades (see Honey 2008) reflects the activity's implicit promise to redress a variety of problems created by processes of capital accumulation. Specifically, ecotourism may provide a spatial fix in its investment of excess capital in new geographic development; a temporal fix in its selling of an ephemeral experience that is instantaneously consumed and must therefore be continually repurchased (hence minimizing turnover time for recovery of invented capital); a time-space fix via foreign lending for ecotourism development (i.e., that offered by the World Bank); a "social" fix in its effort to counteract capitalism's tendency toward inequality by redistributing wealth back to poor rural communities; and "psychological" fixes in the selling of an experience of "nature-culture unity" to redress the common sense of humans' alienation from nonhuman natures wrought by capitalism's metabolic rift (Neves n.d.), as well as an experience of sensation-rich enchantment to counter capitalist modernity's concern with rationalization and disenchantment (Fletcher and Neves 2012). In this sense, as Büscher (2012) describes, capital expansion through ecotourism is framed as a solution to the ecological (and social) contradictions of capitalism itself, promising a form of production and consumption that not only does not degrade but in fact enhances both environments and social relations.

This presentation, however, conceals a number of contradictions intrinsic to the very process of ecotourism development, whereby significant ecological and social costs are obfuscated via a pervasive ecotourism "bubble" (Carrier and Macleod 2005) that emphasizes ecotourism's positive potential and minimizes its negative consequences (Fletcher and Neves 2012). In terms of the Lacanian framework outlined above, this can be seen as an instance of the inevitable gap between Real and Symbolic sutured in the Imaginary via the free-market fantasy that further, intensified application of neoliberal principles can not only fulfill ecotourism's lofty promise but also rectify these evident deficiencies in its implementation.

This obfuscation is aided by a further dimension of ecotourism development that, as intimated earlier, takes the body as its focus. In addition to expanding geographically, in other words, ecotourism transforms the human body itself into an important site of capital accumulation (Harvey 2000), thereby providing what might be called a "bodily" fix (Guthman and DuPuis 2006) to complement the others previously outlined (see Fletcher and Neves 2012). It does so by turning the body *itself* into a site of accumulation (Fletcher and Neves 2012). What ecotourism as a capitalist "product" offers most centrally is an experience of *jouissance*, a feeling of intense sensation or excitement that, although commonly framed as unequivocal enjoyment, in fact frequently entails significant negative emotions and bodily sensations as well, confirming the ambiguous status of *jouissance* as a mixture of pleasure and pain (Kingsbury 2010, 2011).

Even this ambivalent state is fleeting, however, as the sensation recedes and mundane reality intrudes once more. In other words, the affective release offered in ecotourism is transitory, and hence, rather than delivering an enduring satisfaction of existential angst, the experience usually provides merely a "pseudocatharsis" that paradoxically leaves the subject even more dissatisfied through deprivation of the previous stimulation. Yet the fleeting experience provides enough pleasure that its subsequent withdrawal inspires a desire for further experience in the hope of recapturing the previous "high" and thereby achieving the enduring resolution thus far denied. In this way, an opportunity for further accumulation is created as tourists seek to reexperience the desired emotional stimulation in search of an illusory satisfaction. As the object of this process is an ephemeral affective state that passes quickly with little residual impact on the body, this accumulation process can be virtually infinite, facilitating continual capitalization without readily discernible limit or consequence. In this way, Lacan observed, "surplus enjoyment" and "surplus value" go hand in hand, stimulation of the first facilitating accumulation of the second by compelling increased consumption of the products and services through which *jouissance* is pursued (Žižek 1989). In the process, the *jouissance* conferred by the ecotourism experience helps to sustain a global industry of substantial proportions (Honey 2008), the fastest-growing segment of an international tourism market currently valued at more than US$1 trillion (UNWTO 2012).

The *Jouissance* of Ecotourism

My empirical research concerning the practice of ecotourism, conducted primarily through participant observation with whitewater paddlers (both rafters and kayakers) on expeditions ranging from one day to two weeks in a number of locations throughout North, Central, and South America, illustrates this dynamic (see Fletcher 2009). "Serious" paddlers—those who engage in their pursuits independently and regularly, often doing so professionally (for nominal pay provided through equipment sponsorships and suchlike)—are commonly caught up in an incessant quest to experience new rivers in new locations, progressively increasing the difficulty of their endeavors as they grow comfortable with previous challenges that consequently no longer provide sufficient stimulation to be compelling. In this manner, many paddlers spend the bulk of each year traveling from river to river around the globe, expanding, in the process, the ecotourism industry of which they are part and parcel. One of my key informants, for instance, could reasonably be considered an "adrenaline junky," seeming unable to stop engaging in quite perilous undertakings that, indeed, nearly cost him his life on several occasions. In the year that I first encountered him, he spent the summer kayaking and working as a commercial raft guide in California, then moved east for a month to paddle in West Virginia, after which he migrated to eastern Canada to continue raft guiding there. Following this, he spent several weeks in Jamaica pioneering a new commercial raft operation before continuing on to Chile, where I encountered him again in Patagonia, working one of the most difficult rivers in the world. Over the next several months I watched him consistently push the limits of safety, dumping clients into the river on several occasions, including once when he drove his raft straight into a series of monstrous waves that caused him to flip end over end in one of the most spectacular crashes I have ever seen. In the years since, he has continued his adrenaline odyssey, earning recognition for his skillful first descents of several previously unrun rapids while almost losing his life several more times as well.

A similar escalation occurs with clients on commercial whitewater rafting trips. In Chile I participated for several months as a guide on a series of weeklong rafting trips widely considered among the premier ecotours in the world (see Fletcher 2009). For clients, this trip was often the crowning moment of a long history of engagement in similar pursuits, which they took great pleasure in describing as we sat around the campfire during downtime after a day on the river. By the end of the trip, however, many participants were already planning their next excursion, inquiring what I and the other guides thought was "the next most difficult river" after the one they had just completed. Neves describes a similar situation on whale-watching tours (in Fletcher and Neves 2012).

These dynamics are supported by construction of what I call an "ecotourism imaginary" composed of fantasmatic representations of past expeditions—usually tied up with European colonialism (see Braun 2003)—that functions as a model for ecotourists' undertakings, suggesting that the latter can achieve experiences similar to the larger-than-life protagonists of these celebrated tales. Several clients on the Chilean raft trip, for instance, claimed to have been inspired by Shackleton's famous (aborted) Transantarctic Expedition of 1911–12 (see Lansing 1959; Fletcher 2009). The narratives describing these expeditions, however, are of course merely aestheticized condensations of the actual (Real) experiences they reference. Descriptions of colonial expeditions, for instance, commonly whitewash the negative impacts of colonialism and emphasize the ostensive adventurous heroism of the colonists themselves (Fletcher 2012b). Hence, the representations they convey are largely fantasies, motivating not through their realistic portray of the grueling rigor of historical experience but through mobilization of desire for the pleasurable emotion they ostensibly offer.

Conclusion

In dynamics such as this, processes of production and consumption in neoliberal capitalism are synchronized through their shared pursuit of *jouissance*. Fantasies of the pleasure attendant to endless accumulation on the part of producers are supported by fantasies of limitless pleasure promised by sensation-rich experience on the part of consumers, allowing both parties to participate in the construction of a global industry/imaginary. This is further reinforced by market mechanisms' claim not only to avoid negative social and environmental impacts but on the contrary to confer, through both production and consumption, substantial socioenvironmental benefits, thus offering as well the promise of the *jouissance*-laden "warm glow" produced by charitable giving (see Wilson 2012), which Bishop and Green (2008, 39) indeed compare to "the dopamine-mediated euphoria often associated with sex, money, food, and drugs."

This analysis therefore suggests that attachment to an ideology such as neoliberalism is a function not merely of cynical manipulation in the interest of accumulation, disciplinary internalization of power relations, or even simple provision of monetary incentives but also of a visceral, libidinal investment in the pursuit of *jouissance*. In this way, as Wilson (2012, 7) describes, "beneath the articulation of discourses, the organisation of institutions, and the arrangement of social relations, *jouissance* is mobilised and regulated through disavowed social fantasies that structure relations of domination in ways that displace or foreclose their constitutive antagonism." This dynamic may help to explain the remarkable resilience of neoliberal ideology within environmental governance (and beyond) despite neoliberal policies' recurrent inadequacy in actual practice.

How, then, is critique to effectively confront this situation? Acknowledging the implications of the disavowal he highlights, Žižek (1989, 25) cautions that "we must avoid the simple metaphors of demasking, of throwing away the veils which are supposed to hide the naked reality. We can see why Lacan, in his Seminar on *The Ethic of Psychoanalysis*, distances himself from the liberating gesture of saying finally that 'the emperor has no clothes.'" Rather, Žižek suggests, the aim of critique must be "to detect, in a given ideological edifice, the element which represents within it its own impossibility" (143), to undertake a "symptomal reading" that seeks "to discern the unavowed bias of the official text via its ruptures, blanks, and slips" (1997, 10). As Kapoor (2005, 1205) paraphrases, "This means tracking and identifying ideology's Real—its slips, disavowals, contradictions, ambiguities."

For neoliberal conservation, this Real may manifest as symptom in growing evidence concerning the overwhelming failure of more than thirty years of persistent global efforts to integrate conservation and development by harnessing market mechanisms to commodify in situ natural resources and thereby incentivize their preservation (e.g., Fletcher 2012c; McShane et al. 2011; Wells and McShane 2004). What this failure may point to, indeed, is nonhuman nature's essential recalcitrance in the face of persistent attempts to make it pay for itself, the eruption of the unruly "Real of nature" (Stavrakakis 1997a; Swyngedouw 2011) through the neoliberal fantasy that the same capitalist forces that commonly exacerbate both poverty and ecological destruction can be employed to resolve these selfsame problems. Stavrakakis (1997a, 124), for instance, suggests, "What is shown by the current environmental crisis . . . is that there are in fact some limits, limits to growth and economic expansion, limits imposed by the Real of nature." Highlighting the mechanisms sustaining disavowal of this realization, our urgent task is to "accept fully . . . the Real in its irreducible constitutivity" (128). As Büscher and colleagues (2012) advocate, this entails working to develop (as well as recognize and support where they already occur—see Sullivan, this volume) alternate modes of being that do not commit such violence, either physical or epistemic, to the "more-than-human" world in which we live.

CHAPTER FIVE

Celebrity Spectacle, Post-Democratic Politics, and Nature™ Inc.

Dan Brockington

Two frontiers of capitalist expansion are restructuring environmental resource use and policies. There is a new round of landgrabbing by elites, corporations, and governments worldwide that is threatening the resource bases of the rural poor (Borras et al. 2011; Fairhead, Leach, and Scoones 2012b). And there is the reconfiguration of environmental policies around payments for environmental services, which again could substantially compromise rural livelihoods (Engel, Pagiola, and Wunder 2008; Redford and Adams 2009; Kosoy and Corbera 2010; Brockington 2011). Both trends are driven by the same dynamic—the relentless capitalist pursuit of profit, which seeks to enclose new lands and create (and take possession of) new commodities.

In such a context, spending any time thinking or writing about celebrity may seem foolish. Barricades and protest are surely more intelligent responses to landgrabbing than reaching for the nearest copy of *Heat* (a celebrity news magazine). I have some sympathy with that response; I have firsthand experience of how useless *Heat* can be when opposing landgrabs. However, if we follow the politics of celebrity endorsements for environmental issues and other good causes, then we can learn a great deal about how the political space and policy environments have been created that enable such appropriations to take place.[1]

To understand the nature of celebrity's role in Nature™ Inc. we must make three routine observations. First, there is not enough critical public or academic debate about these twin capitalist frontiers. As Redford and

Adams (2009) have complained, payments for environmental services are greeted too enthusiastically. Landgrabs are portrayed as attempts to solve global food shortages or replace hydrocarbons with green biofuels. The persuasive forces of profit-seeking capitalism, with decades of experience of guile behind them, and the marketing branches of NGOs are being directed at selling these possibilities. Accumulation through dispossession can be welcomed as a progressive move.

Second, we must note that celebrities are part of the persuasion to support these moves. The Prince's Trust (established by Prince Charles, first in line to the British throne), for example, has been adept at enrolling a large number of international names to promote plans to make the carbon in rain forests worth enough money for capitalism to seek to preserve it (Igoe 2010).[2] Harrison Ford (an actor) supports Conservation International to advertise the possibilities of realizing value from ecosystem services in Africa. Even if Ford somewhat gauchely declared that Africa was a continent where "nature and people are one," the message was clear. Nature is good for business, and African ecosystems are a good place to conduct it.

Finally, we must note that celebrities are also part of the opposition. Consider, for example, the appearance of Paul McCartney (a musician), Robert Redford (an actor), and Cilla Black (a TV personality) joining Greenpeace to speak out against oil exploration in the Arctic during the recent Earth Summit.[3] They were initiating a signature petition that will enroll more than a million signatures and that will then be deposited symbolically beneath the ice of the North Pole. Their stance provides an apparently welcome contrast to more establishment-driven publicity. It surely represents an instance of popular and civic power taking on political and economic power in opposition to capitalist expansion.

In some respects their stance is just that: it both seeks to limit capitalism's terrain and may create the political space for more opposition. But we cannot just take this campaign at face value. Look at the sorts of politics that this is promoting. Burying signatures beneath the North Pole creates good spectacle, but it does not give those signatories much voice or visibility. Rather, it gives Greenpeace political capital to wield; the primary direct beneficiary of this Arctic petition will be Greenpeace and not the Arctic. This is the same organization whose support of the newly declared Chagos Marine Protected Area has aroused some controversy because that support did not, initially, make clear enough Greenpeace's backing of the claims of formerly resident Chagossians or their opposition to the presence of a massive, polluting American military base on Diego Garcia in an exclave in the heart of the park (Sand 2010). This unusual example shows that

mainstream conservation organizations can be rather useful to the establishment. Celebrity opposition to capitalist expansion, therefore, needs to
be taken not just at face value but in terms of the broader relationships that
celebrities signify and empower.

We need to see the very presence of celebrity voices as a testimony to the
incorporation of NGOs and civil society groups by mainstream politics and
businesses. At the very moment when celebrities protest capitalist expansion, they are demonstrating and furthering its power. The marchers against
capitalism are also its vanguard, for the incorporation of celebrity, and the
cultivation of relationships that make petitions like Greenpeace's possible,
reflects a reorientation and reconfiguration of the corporate sector. This
reconfiguration has seen closer and more intimate relationships develop
between NGOs (also referred to here as the "third sector"), the celebrity
industries, and the corporate sector.

The significance of this alliance is best understood in terms of Colin
Crouch's (2004) idea of "post-democracy." Post-democracies are characterized by disengagement and apathy with respect to politics by much of the
citizenry that democracies are meant to empower. They are characterized
by increasing inequality but apparent popular acquiescence to this fact.
Politics and government become the domain of elites, lobbyists, and special
advisors. These societies are liberal in that they allow free rein to all sorts of
diverse voices and movements. At the same time, these lobbies and advisors
are particularly vulnerable to domination by the wealthy, who have the
most resources to promote their interests. For Crouch (2004, 13, 52), the
power of corporate elites is the "fundamental problem" of postindustrial
societies, because one of the "core political objectives of corporate elites
is clearly to combat egalitarianism." He finds that the post-democratic
societies have been stricken by the "establishment of a new dominant,
combined political and economic class" (52) that mirrors far too closely
the pre-democratic politics of the nineteenth century.

This is why celebrity matters. Vast lands are being sucked into the corporate maw. Nature's intimate inner workings are separated, packaged,
commodified, and sold. And the political space that makes these policies
possible, the thinking that advances them, and the interest groups that
lobby for them are becoming the territory of elites amongst whom celebrities play a prominent role. Moreover, celebrity is an important part of the
circulation of images and ideas upon which the creation of new forms of
speculative, and often spectacular, value can hinge. Celebrity is not just
ornamental to this process, it is constitutive of it.

I will demonstrate below more comprehensively that celebrity is part of Crouch's political elites, but to do so I need to extend his argument in one respect. While Crouch notes the rise of issue-based politics and lobbying organizations, he does not make clear enough the role of NGOs in constructing and speaking with the political elites. As I show below, Corson's (2010) accounts of the International Conservation Caucus Foundation or Holmes's (2010) exploration of the transnational conservation class or Sireau's (2008) account of Make Poverty History will quickly underline how important these groups are in elite formation and reproduction. By following the work of celebrity in NGOs, we will quickly scent both the power and reach of corporate influence within the third sector, as well as the importance of the role of celebrity lobbying among these elite groups.

These arguments will also require me to extend the normal subject matter of this volume and talk not just about environmental NGOs and environmental politics but about the NGO movement as a whole, with a particular focus on development issues. I can defend this on three grounds. First, while environmental NGOs are extremely powerful and influential in themselves, their dynamics have to be understood within the context of broader changes across NGOs. The division between development and environmental NGOs often breaks down anyway. Environmental NGOs take on development projects and vice versa. Second, I have a wealth of interview material from across the environmental and development organizations in the third sector that demonstrates the need to consider them together.[4] Third, environmental issues, particularly within poorer parts of the world, are simply an aspect of development issues. Decisions to conserve a national park are part of development plans. Payments for environmental services are a development project, they are part of immanent and planned development. We have as much to learn about Nature™ Inc.'s promulgation from the activities of development NGOs as we do from specific environmental NGOs.

The argument proceeds as follows. I examine how celebrity and the third sector have become more intertwined over time and how that process reflects and is driven by the strength of corporate interests. I then examine the role of those interests in contemporary capitalist conservation with respect to the creation of elites and with respect to the creation value through circulating images. The role of celebrity in the construction of potentially regressive elites is not, however, the end of the story. I will also argue that celebrity involvement can advance radical causes. In the conclusion I offer my reasons for finding some radical potential for celebrity activism.

Celebrity, NGOs, and the Corporate Sector

There has been a shift across the third sector to embrace more effectively the opportunities that celebrity is perceived to offer. The shift takes the form of a more systematic, organized, and professional approach to relations between NGOs and the celebrity industries. The indications are numerous. Increasing numbers of NGOs (75 percent of the largest thirty in the UK, according to a BBC report) now have dedicated celebrity liaison officers whose job it is to manage relations with high-profile personalities. In the UK the Media Trust, which promotes effective media use by charities, runs bespoke celebrity liaison workshops for NGOs who want to learn how to work with celebrities. NGOs sign up to professional celebrity contact databases, which keep updated lists of how to contact different public figures—and these databases advertise their more important NGO clients to signify the value of their services. The Red Pages, for example, announces its services to Oxfam and the British Red Cross on its front page, along with other commercial clients such as M&C Saatchi and Vivienne Westwood.[5]

I date the professionalization as occurring in the UK mostly since 2000, as most celebrity liaison within NGOs seems to have begun full time and in earnest since then. The celebrity liaison officers' forum, a monthly meeting, has been convening in London for about eight years. But there are precursors—Oxfam's first such appointment was in 1994. It probably has a longer history in the United States, where, for example, the comedian Danny Thomas was particularly effective at mobilizing celebrity support for the St. Jude's Children's Research Hospital since the 1960s (Weberling 2010) and where the Elizabeth Glazer Pediatric Aids Fund took a lead in promoting celebrity support for HIV/AIDS research.

The professionalization of relationships between NGOs and the celebrity industries did not initiate an increased wave of celebrity-focused publicity. It is an evolution in response to steadily increasing needs of association with celebrity. This much is plain from trends in the increase in articles about charity that mention celebrity. Figure 5.1 presents trends from the *Guardian* newspaper, for which we have the longest records. This shows a steady increase up to about 2005, when, for reasons that remain unclear, the persistent rise begins to slow, if not decline. Note too that the proportion of charity articles that mention celebrity is greater than the proportion of all articles that mention celebrity. In other words, the charitable sector has been more enthusiastic in its embrace of celebrity than have other sectors of society.

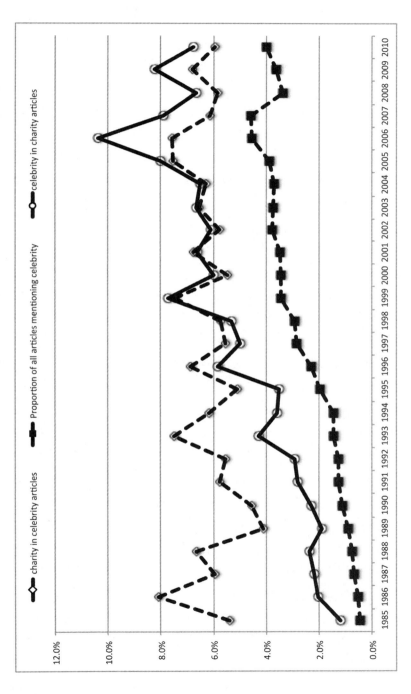

Figure 5.1. Trends of mention of celebrity and charity in the *Guardian* newspaper.

If we reverse our gaze and examine the response of the celebrity industries to the charitable sector, then a different pattern emerges. In brief, the proportion of articles about celebrity that mention charity has increased only marginally. Most celebrity articles are still about the traditional fare of celebrity. The rise of "charitainment" (as *Time* magazine heralded the celebrity-drenched fare of 2005) has not substantially affected reporting about celebrity matters (see Poniewozik 2005).

There has, however, been a response from the celebrity industries to the opportunities of the charitable sector. Three of the four major talent agencies in Hollywood have, since 2000, established foundations that manage their corporate social responsibility interests and, crucially, the charitable interests of their clients. Dedicated staff facilitate their clients' expression of their charitable interests, providing them with contacts, reading, and information on the causes that they wish to pursue. As Rene Jones, who provides this service for clients of the United Talent Agency, explained to Jonathan Foreman (2009), "It's mostly counseling advice. You meet, get to know one another, and then you act as a matchmaker and bring credible organisations to their attention. It's not always an instant process, but the ones that last longest are those with an organic connection."

There are also specialist independent companies that act as brokers between wealthy stars and the causes that they wish to pursue. One of the more prominent is the Global Philanthropy Group, whose clients include prominent Hollywood A-listers and whose investigation into the musician Madonna's charity Raising Malawi led to a rather damning report and his company taking on Madonna as a client.

Finally, we should note that speaking out for good causes has become a substantial part of some celebrities' brands. Bono (a musician) is so well associated with poverty relief in Africa after his work for Band Aid, Live Aid, the Jubilee Debt Relief Campaign, DATA, ONE, the G8 of 2005, and Live 8 that he has appeared in Louis Vuitton advertisements with the title "All the best journeys begin in Africa." Angelina Joli (an actress), who was also featured in this series, was photographed in Thailand. Similarly, George Clooney, Don Cheadle, and Matt Damon (all actors) are known for their interventions in Sudan and in Damon's case with water development and microfinance. With respect to the environment, the musician Sting has maintained a persistent interest in rain forest conservation, and there is a persistent trickle of nature documentary presenters who, achieving fame as nature's spokespeople, also support environmental charities.[6]

The growth of celebrity in charities, despite its many potential financial and political advantages for NGOs, is not at all something that the sector is responding easily or happily to. Rather, there is often a lingering suspicion about the inequalities and excess of celebrity. Some NGOs deliberately eschew any form of engagement with celebrities. Part of their brand and politics is to work without them. Even in NGOs that do work with celebrities, their celebrity liaison officers can feel marginalized within the organizations. The origins of the celebrity liaison officers' forum was in the chance meeting of officers who were delighted to meet colleagues who shared their difficulties and problems at work. One of the most prominent and distinguished environmental NGOs has only recently begun organizing and investing seriously in formal celebrity liaison operations. I have it on good authority that the chief communications officer of another major environmental NGO was unable to identify a picture of Madonna's face (she has long been extremely famous).

Even if the trend is receiving only a cautious and equivocal welcome from some members and employees of NGOs, it is nonetheless an increasingly important aspect of the way that the third sector is working. As we have seen, publicity and fundraising events can often be strongly flavored by celebrity. Examining some of the reasons behind this shift will help us to consider its consequences.

The reasons for this shift in the NGO sector are multiple. We must not ignore the importance of individual sentiment, especially for those celebrities who are more prominently involved. But we also have to recognize the numerous structural factors driving the growth of these interactions. As I have argued previously (Brockington 2009), part of that growth is simply due to the increasing numbers of NGOs and celebrities over the last twenty-five years. Their mutual need for good publicity has resulted in more interactions.

However, for the purposes of this chapter, and particularly the argument I advance with respect to post-democracy and the dominance of corporate interests, we need to recognize another driving force. The interactions of celebrity and charity are fueled by a strong corporate interest in getting access to celebrity. Across the interviews I conducted, the corporate fascination with celebrity proved a constant theme. Corporates are "starstruck," and they "really liked having celebrities involved" (Sources 35 and 48). The corporate teams within NGOs can be those who make the most requests for celebrity involvement.

Celebrity has long been a significant aspect of corporate advertising. Getting the match of celebrity with corporate brand right is a matter of keen investigation by a number of researchers, for some associations have been astonishingly successful and lucrative (Pringle 2004). But "it's very expensive for a corporation to get a celebrity spokesperson, so they love it when by doing good works they also get to grab a few photos . . . or do a joint press conference with a celebrity. It's a big added benefit. . . . They love getting that opportunity to be associated with a celebrity for free" (Source 55).

One of the attractions of working with charities with good celebrity contacts is that it provides the opportunity for celebrities to be associated with these companies' products and brand. Charities have to be extremely careful that they do not let their celebrity spokespeople become free endorsers of corporate products, as that would threaten the celebrity business model. More than one organization has drawn up written guidelines for celebrity liaison officers to follow so that they do not cross the line beyond a legitimate partnership between celebrity, charity, and corporation.

But the dividing line between what can be done freely for the charity and what must be paid for by the corporation can be thinly drawn. Public figures can endorse the relationship that companies may have with NGOs, but they may not endorse the company itself. Furthermore, some associations with particularly well known charitable causes can result in further opportunities for (paid) corporate endorsement for the celebrity. Some public figures take the resulting work and endorsement opportunities. Others refuse to take on such endorsements, saying, "It is an insult to be offered to be paid" as a result of any association that arises out of work for charity (Source 71). Numerous charities for their part advertise the possibilities of association with their celebrity spokespeople on their corporate webpages. They seek to recommend their services to potential corporate partners in terms of the celebrities with whom they are associated.

It is not just the possibilities of free access to expensive people that appeals to corporate partners. Sometimes there are strong brand considerations at work. Certain NGOs can host prestigious and plush events with celebrity attendees with whom it is simply good business to be associated. It is important to appreciate the scale of these events. Journalists from the *Daily Mirror* have recently compiled a league table of the most expensive and luxurious such occasions, a table that bears summarizing (see table 5.1), because the expense and luxury of these events is simply mind-boggling. It is not surprising, therefore, that some corporate sponsors are so keen to be associated with these events: "In the first instance the draw is going to be that to have [public figure E] associated with their product is highly desirable, and [then] to sponsor an event that we do at [location F], which is very elite [and] which an awful lot of very rich or famous people come to, is perfect to their brand" (Source 67). In some instances the importance of corporate sponsorship is such that it is the corporate interests

Table 5.1. Costly charitable occasions

Event	Year	Cost (£)	Money Raised (£)	Costs as a proportion of money raised
Sunseeker Charitable Trust Ball	2009	400,000	550,000	72%
Raisa Gorbachev Foundation Gala	2007	400,000	1,200,000	33%
Prince's Trust Berkeley Square Ball	2008	420,000	450,000	93%
Caudwell Children Annual Ball	2009	818,000	2,520,000	32%
Grant and Anthea's Summer Ball	2007	829,000	942,000	88%
Elton John's White Tie and Tiara Ball	2007	972,000	8,010,000	12%
ARK Gala Dinner*	2007	3,900,000	26,000,000	15%

Source: Sommerlad 2009.

* The large sums here are accurate. The dinner is hosted by a billionaire financier who insists that the party must be "mind-boggling" for his guests, who pay £10,000 for a ticket. The organization has also tried to curb the expenses of subsequent galas, frugally keeping them to below £2 million for an evening.

more than anything else that determine the level of celebrity engagement. One charitable campaign reviewed its activities and questioned whether it should continue working with celebrities and instead use "real" women in its public announcements. The campaign realized that fewer celebrities might be better for some of the consumers of its advertisements and messaging, but "we came to the conclusion that actually the corporates like to be associated with the campaign because of the kudos it holds and because of the level of celebrities it has supporting it. Because the majority of the income is raised through corporate partners . . . we do actually need to have celebrity support to keep them engaged" (Source 63).

Finally, we must note too that the corporate fascination with celebrity sponsors at charitable events is not just about dismal economics and the bottom line. It can include other, more personal inclinations. One of the pleasures and privileges of success in business is that one gets to meet in person the famous people everyone else only hears about in the media: "[Company G] absolutely loved [public figure H]. . . . [S]he's very special to a group of men of a certain age. . . . She's . . . lovely and charming and articulate. . . . [S]he delivers both emotionally and rationally and . . . appeals to the softer side of them" (Source 48). Similarly, politicians are most aware of the beneficial publicity that can result from being seen to be associated with popular public figures. They are keen to meet them and to be seen to meet with them: "You might be able to meet with someone lower down in the office, but suddenly you are meeting with the chief of staff or with the principal instead of a staff member two or three levels below because you are accompanied by a celebrity. You also might be able to get a hearing on Capitol Hill because one of those testifying would be a celebrity. . . . That happens all the time" (Source 93).

Celebrity, then, is not just a pawn in negotiating arrangements between charities and corporations, nor is it just a vehicle for getting into the news. It is also a lubricant in the negotiation machinery, it helps bring people to meetings, it facilitates the negotiation of deals, and it enables a large number of policy and financial discussions to take place at a speed and with a conviviality that would not otherwise be possible.

These driving forces present two important areas of investigation for understanding the implications of celebrity for the evolution of Nature™ Inc. These are, first, the role of celebrity in facilitating the production and reproduction of an elite whose ideas are shaping the introduction of new environmental discourses, and, second, the role of celebrity in the circulation of images and representations of nature on which so much environmental policy hinges. We will deal with each in turn below.

Evolving Elites and Circulating Celebrity Images

The construction of conservation and environmental elites has already been well studied. Perhaps the most specific contribution is George Holmes's (2010) work on the creation of a "transnational conservation class." Drawing on Leslie Sklair's (2001) ideas of a transnational capitalist class, Holmes postulates the existence of a similar transnational group of elite conservationists comprising scientists, bureaucrats, NGO workers, and government officials as well as environmental celebrities. In his own work he demonstrates that these elites can sometimes function in contrary ways—they are nationally inspired at the same time as they are transnational in their socializing, such that, in his case, elites from the Dominican Republic resisted the influence of US elites.

There are numerous other studies. Catherine Corson's (2010) examination of the International Conservation Caucus Foundation provides a good indication of the way that elites function and the role of celebrity in providing a hinge and focus to plush galas that bring leading conservation NGOs into pleasant meetings with prominent politicians and business leaders. One of the first such meetings honored Harrison Ford with an award for his support for conservation. Another example comes from Ken MacDonald's (2010a, 2010b) investigations (and see MacDonald and Corson 2012) of the role of royalty and other celebrity elites in the construction of hegemonic conservation spectacles at international conservation meetings.

It is plain that here the role of celebrity is to provide a theatrical focal point (e.g., in the form of prize givings) to meetings that are mainly about facilitating contacts and agreements and a whole host of other interactions taking place in the wings. They enable collective endorsements to be performed at large gatherings, as MacDonald has described, producing consent and support for the prophets and leaders of the new environmental policies. Celebrity, not surprisingly, is integral to the establishment and reinforcing of hegemony within environmental movements.

Researching such gatherings and networks is notoriously difficult—we simply cannot get access or close enough to the actual proceedings to learn what is going on. We have instead to rely on historical accounts, when archival records become available, or to study their effects from a distance. Occasionally, however, we can get glimpses behind the scenes at what people are saying from the odd report into these privileged gatherings.

One such is available from a gathering of the Brookings Blum Roundtable in 2007, which brought together a number of famous faces and people close to famous faces in the development and environmental fields (Brainard and LaFleur 2007). Attendees at the three-day event included former US vice president Al Gore, former Irish president Mary Robinson, former US secretary of state Madeleine Albright, cofounder of Product (RED) Bobby Shriver, and World Bank managing director Ngozi Okonjo-Iweala. It was, in other words, a high-level gathering. Its purpose was to look at the new actors in development and the environment, including celebrities, philanthropists, and the private sector. The report of that gathering (Brainard and LaFleur 2007) must be treated with some caution, as it is a public-facing document that is unlikely to report the disagreements and debates that are bound to have animated the discussion. Nonetheless, precisely what is interesting about it is what the participants and rapporteurs were prepared to agree about in public, especially as far as celebrity is concerned.

The overwhelming tone when celebrity is mentioned is of approval and commendation. Celebrities are "maximizing the power of their public appeal to champion global poverty awareness and activism" and "injecting a dose of credibility and charisma into the foreign assistance and development debate" (Brainard and LaFleur 2007, 4). They "have focused public attention on humanitarian crises such as HIV/AIDS and the conflict in Darfur" (5) and helped to raise billions of dollars in development funds. Perhaps most of all, they are infusing some passion, joie de vivre, anger, and excitement into topics that can be dull to Western minds:

> Whether rock stars, movie stars, moral leaders, or political icons, these "celanthropists" are infusing antipoverty campaigns with their own charisma and brand allure. Some are adept at crystallizing complex issues in catchy slogans like "Drop the Debt" and "Make Poverty History." Others have made energetic use of the popular media to attract new development audiences; witness MTV's Diary of Angelina Jolie and Dr. Jeffrey Sachs in Africa. Seasoned performers on the global stage, these development champions are eloquent and impassioned in their appeals on behalf of the impoverished—invoking emotional language and images designed to anger, engage, and inspire action. (Brainard and LaFleur 2007, 16)

The report even finds a way of welcoming the fact that these new activists are inexperienced but still believe they can make a huge difference. This sort of thing is normally called arrogance, but here this trait is welcomed thus: "Many of the new development players are entering the field unburdened by the weight of conventional wisdom and are blessed with confidence in their own ability to achieve outsized results" (Brainard and LaFleur 2007, 35). These representations are interesting partly because of what is missing. For example, the role of celebrities in the Make Poverty History campaign is simply and only painted in a positive light, even though it caused the most intense controversies at the time (Sireau 2008). The achievements of celebrity interventions in Darfur, again reported favorably, are also hotly disputed (Flint and De Waal 2008; Crilly 2010; Hamilton 2011). The report does mention problems of simplification of messages and refers to the uneasy relations that can exist between the two celebrity industries and NGOs, but these occupy just three paragraphs in a long document.

The report is also interesting because one of the strongest characteristics of elite approaches to celebrity, at least as identified in this report, is their faith in the power of celebrity, particularly with respect to converting the public to their causes. This is stated in the baldest terms: "And it works. The public is answering their call in unprecedented numbers" (Brainard and LaFleur 2007, 16). Al Gore is quoted as saying that "rock stars proved instrumental in supercharging a new generation of climate crusaders during 2007's Live Earth" (18). But what sort of public voice, produced by these celebrity interventions, is being welcomed by these elites? This is one of the most significant aspects of this report. For what this elite gathering celebrated as examples of the public voice are "the hundreds of thousands who attended the ten 'Live 8' concerts in the run-up to the Gleneagles summit, the more than 2.4 million signatures for the ONE Campaign, and the 63.5-million-strong audience for the 2007 U.S. television special American Idol: Idol Gives Back" (6). In other words, what this elite gathering celebrated were rather passive audiences expressing themselves through attending concerts, texting donations, switching on televisions, and signing on-line petitions. The marches of Make Poverty History, so controversially eclipsed by the Live 8 concerts, do not even get a mention. The very signs of democratic disillusion that so worry some commentators are welcomed here as evidence of success.

This document celebrates the power of celebrities to concentrate their voices into the hands of organizations that can then represent their interests at elite gatherings. While the elites at the Brookings Blum roundtable were probably among the most genial and egalitarian possible, their enthusiasm is for the elite-privileging politics of post-democracies. Indeed, it would be difficult to find a clearer example of Crouch's (2004, 19–20) depiction of post-democratic politics and public roles: "The idea of post-democracy helps us describe situations when boredom, frustration and disillusion have settled in after a democratic moment; when powerful minority interests have become far more active than the mass of ordinary people in making the political system work for them; where political elites have learned to manage and manipulate popular demands; where people have to be persuaded to vote by top-down publicity campaigns."

To understand the role of celebrity in circulation I draw on Büscher's writings about the importance of circulation for our understanding of how value can be created out of conserved nature (see Büscher's chapter in this volume). Büscher notes that capital can seek not to transform nature but to conserve it (e.g., in sequestered carbon or offset biodiversity) in order to create value from it. In such circumstances he argues that value is socially produced through expert assessment and a host of practices that "make 'liquid nature' believable, legitimate, and manageable," to achieve which "capital has had to and continues to create particular governmentalities and associated ideological belief systems." The crucial result of all this labor is that in "turning conserved nature into capital, conservation has become fictitious; it can still sell. All that it needs is a compelling brand: a memorable logo, some catchy slogans, smooth marketing campaigns, visually captivating websites, celebrity spokespeople, and a take-home message that 'everybody wins.'"

Celebrity features prominently here, as Büscher notes in his chapter. But it is important to note how closely and well celebrity fits with requirements of this circulating nature. First, celebrity exists as circulation. That is its very essence. Celebrity is about being seen and noticed as often as possible and in as many different contexts as possible. Indeed, the rise in celebrity over the last twenty years partly reflects the changing conditions of circulating images within the media. As media outlets have been taken over by giant companies that have extended their control vertically and horizontally, so it becomes even more profitable for different companies owned by the same company to promote celebrities featured in each other's news (Turner 2004). Their images circulate across different outputs. Thus we find Lily Allen (a singer) promoting rain forest conservation for the satellite television company BSkyeB in conjunction with the WWF. The company is seeking to act thus because such charitable activities are good for its brand: consumers like it. And the trip is covered by newspapers that are also owned (as is BSkyeB) by Rupert Murdoch, whose journalists were flown out, at Skye's expense, to write the story.[7]

But celebrity also matters particularly with respect to the character of the changes to the economy and society that Nature™ Inc. *envisages*. For, in my assessment, the payments for ecosystem services, natural derivatives, and other new forms of financialization are more grandiosely imagined and vigorously projected than they are actually achieved.[8] And in this realm of possibility and promise, dream and expectation, celebrities, who are people who exist as objects of desire and expectation, come to the fore. As Gabler (1998, 51) put it when analyzing the power of movies in early twentieth-century America: "The America of rapid industrialization, urbanization and immigration, was suffused with a new sense of possibility that made its citizens especially susceptible to the movies' fantasies." By analogy, as ecologists, conservationists, and entrepreneurs flock to payments for environmental services, biodiversity offsets, and other forms of incorporated nature, the same attraction of possibilities is at work. Associations with celebrity are part of the way in which promise and expectation are built up.

Celebrity also functions to render intelligible economic policies and mechanisms that are complex and hard to understand. As Jim Igoe (2013b) has observed, "Statistics, derivatives, and the like are largely the purview of experts and thus inaccessible. . . . Spectacle, however, can retransform these intangible presentations into compellingly tangible ones designed to inspire the confidence and fire the imagination of policymakers, investors, and consumers."

Celebrities, therefore, matter with respect to circulation because of the tone of the messages and images they circulate. If you want to imagine new commercial opportunities from new ecosystem services, then who better than Richard Branson to endorse them. He's even finding ways of finding money from space travel (and it is precisely the point that these space adventures are still only *proposals*). For Nature™ Inc. to become a reality it will have to spark the interest and imagination of investors, and it helps to enroll celebrity on these projects. For the financial sector believes in celebrities; that sector is part of the elites who do so. Indeed, it has been documented that the very *prospect* of celebrity endorsement tends to meet with approval in the financial markets (Agrawal and Kamakura 1995). Again, note that these are merely announcements of future endorsements, not evaluations of what celebrities do to the product being shifted or to brand awareness. These crucial elites believe that celebrity works. The presence of celebrity at the dawn of Nature™ Inc. will help create this new era of commodified environments.

Conclusion

I hope it is clear from this chapter that celebrity is being increasingly closely integrated into the workings and practices of NGOs and is a vehicle by which they forge closer relations with the corporate sector. I hope it is clear too that in enquiring into the sort of work that celebrity achieves, we can learn more about the politics at work in inaugurating the new nature with respect to both the workings of elites and the mobilization of publics. The presence of celebrity in environmental issues is cause and consequence of post-democratic politics. Finally, I have stressed the role of celebrity in the value-creating circulation of images and possibilities upon which the new economic possibilities of nature depend.

The dangers and deceit of celebrity spectacle should be plain. Igoe (2013b) expresses it most clearly: "This misrecognition of decontextualized circulating objects for ecological connection is perhaps the greatest challenge facing global environmentalism today, precisely because it hinders people from recognizing the types of relationships in which they are actually enmeshed." To the extent that celebrity is implicated in such misconceptions, misdirections, and misleadings, to that extent it is unwelcome. And, as I have previously documented (Brockington 2009), the sorts of nature, environments, and landscapes that attract celebrity can be distorted and misconceived. Celebrity, with its tendency to avoid divisive politics (Meyer and Gamson 1995), is a poor vehicle for communicating the tensions and divisions that must attend any environmental initiative. It will be a poor vehicle for communicating the problems of Nature™ Inc.

And yet I also think we need to be alert to the radical potential of celebrity. This may seem absurd given the inequalities and industries that constitute celebrity. But associations and complicity with capitalism implicate us all to differing degrees. Specifically, members of academic elites, including authors of this volume, are implicated by virtue of whom we educate, who funds our institutions, who profits from our publications, and how much we want to influence policy. We are often just as bound up in the creation and advancement of capitalism as celebrity elites. We just do so with less glamour and style to smaller audiences, at fewer parties, and for less money. And yet academic radicalism is often welcomed as a constructive contribution. I think we need a similarly generous understanding of the potential of celebrity politics.

We need to recognize, as Crouch does, that post-democratic politics are too entrenched to be easily fought. We will not be able to start with untainted institutional forms. So when Crouch (2004, 120) argues that "democratic politics therefore needs a vigorous, chaotic and noisy context of movements and groups [which] are the seedbeds of democratic vitality," these movements are likely to include celebrity advocates. When he recognizes too that new social identities will form and make demands upon democracies that cannot be (at least initially) easily accommodated by existing elites (116), celebrity is likely to feature in the construction of these identities.

I find unlikely support for my argument in Andy Merrifield's book *Magical Marxism* (2011). Given that Merrifield is so inspired by a collective search for authenticity in reaction to the deceptions and false promises of capitalism, given that he draws so heavily on Guy Debord and his hostility to capitalist representations, and given that he deplores the invasion of capitalism into leisure time (with celebrity as the vanguard), using his work to endorse the possibility of radical celebrity might seem strange. But Merrifield's (2011, 18–19) repeated call is for Marxists to recognize the circumstances in which people live now, not to look for workers' collectives and class action when the shop floor is no longer a site of activism or consciousness or even collective experience. Mirroring Crouch, he observes that solidarity is to be found in all sorts of new collectivities (Merrifield lists many such on pages 43–44). He insists that what we need is a new and better fantasy (18) and inspiring dreams (42). Engrained in Merrifield's work is a refreshing determination to find possibilities of the radical in unlikely places and alliances.

And although he does not specify it, these collectivities and the motivations of these new fantasies might include the famous. They are, after all, a rather broad category of people and should include at least some radicals. Moreover, it is possible to imagine his call to "disrupt and reinvent, to create desire and inspire hope" (Merrifield 2011, 18) in a way that could involve collusion with celebrity. Networks of activists (90) might include activist celebrities, if not celebrity activists. This subversive celebrity content will not come from the modalities of celebrity engagements that we know now, but it could emerge from other forms of engagement.

There are three general principles behind these points. First, consumer use of celebrity (which I have not explored here) is too diverse and unpredictable to be determined by its producers. It can be used subversively.

Second, there is a great diversity of celebrity form. In the antics of the Yes Men, for example, fame is put to subversive ends. This is no commercial operation, but nonetheless it depends on well-known staged spectacles, filmed and circulated on the Internet, to poke holes in the glowing self-images of corporate endeavor. And if we admit the radical credentials of this group, then how far up the continuum of fame would we be willing to go?

Third, the aesthetic, Eagleton (1990, 3) insists, cannot be analyzed merely as an instrument of domination or rebellion but must be treated dialectically, for it is inescapably bound up in both: it is "an eminently contradictory phenomenon." Creative genius and the commercial fabric around it can be, and often has been, put to the service of hegemonic regimes (it is partly what makes them hegemonic). But we have to at least be open to the possibility that where celebrities and their audiences are part of a hegemonic regime, this may be something both may wish to free themselves from.

In conclusion, there is good reason to be suspicious and wary of celebrity politics in environmental affairs. It is part of the persuasive armory of corporate power and the construction of remote elites who shape society to their own interests. And yet I will not be surprised if, amongst the vanguard of opposition that is forming in the wake of environmental commodities and landgrabbing, we should find some famous names using their renown effectively for just causes.

Notes

1. I refer readers who are new to the analysis of celebrity literature to these excellent accounts: Marshall (1997); Gamson (2000); Turner, Bonner, and Marshall (2000); Rojek (2001); Turner (2004); and Ferris (2007).

2. I follow in this chapter my practice of explaining why readers might be expected to have heard of any of the famous people mentioned, as we cannot assume that anyone is so famous that everyone has heard of them.

3. See http://www.greenpeace.org.uk/media/press-releases/stars-launch-save-arctic-campaign-20120621, accessed August 25, 2012.

4. This chapter draws upon research undertaken for an ESRC Fellowship (070–27–0035) that I held between September 2010 and September 2012 and during the course of which I conducted over one hundred interviews with members of diverse NGOs, journalists, government officials, and employees in the media industry. These methods are described elsewhere (Brockington 2014). Suffice it to say here that quotations and source numbers refer to these anonymized interviews. An earlier version of the arguments drawing on these interviews was written for and circulated to these interviewees. The feedback I received from them was strongly positive, indicating that they recognized the patterns I had observed in the interview data in their own lives and experience.

5. See http://www.theredpages.co.uk/, accessed August 25, 2012.

6. For a more detailed typology and examples of the work of the different celebrity environmentalists and environmental celebrities, see Brockington (2008, 2009).

7. Source 33 and http://www.thesun.co.uk/sol/homepage/news/Green/2970915/Lily-Allen-backs-bid-to-save-rainforest.html, accessed August 25, 2012.

8. As Thrift and others have insisted, mundane revenue streams have to underlie, at least in the first instance, the fantastic derivatives that were later conjured out of, for example, mortgages (Leyshon and Thrift 2007). I have argued earlier that these mundane revenue streams have yet to materialize with respect to natural capital (Brockington and Duffy 2010b). Using an impressive array of examples and a slightly broader definition of financialization, Sian Sullivan (2012b) has contested that point and demonstrated a number of ways in which revenue streams are forged. I do not think that this disagreement matters much for the argument I am making here. However thoroughly it has been realized or not, the point is that there is still great excitement about the potential of what can be achieved.

CHAPTER SIX

Capitalizing Conservation/ Development

Dissimulation, Misrecognition, and the Erasure of Power

Peter R. Wilshusen

> *Capital is not a thing but a process in which money is perpetually sent in search of more money.*
> — DAVID HARVEY, *The Enigma of Capital*

> *Capital is not a simple relation, but a process, in whose various movements it is always capital. . . . However, as representative of the general form of wealth—money—capital is the endless and limitless drive to go beyond its limiting barrier.*
> — KARL MARX, *Grundrisse*

Over the past twenty years, global conservation efforts have unfolded within the context of two macrotrends: attempts to frame and promote sustainable development and the rise of neoliberalism. More recent initiatives have explicitly joined these two domains of theory and action via global-scale programs aimed at constructing a "green economy." Two high-profile international conferences staged during 2012—the United Nations Conference on Sustainable Development (UNCSD, or Rio+20) and the World Conservation Congress—illustrate how neoliberal capitalism has merged with conservation/development in practice. The Rio+20 meetings, for example, highlighted "green economy in the context of sustainable development and poverty eradication" as one of two overarching themes within a broad statement issued by UN member states entitled "The Future We Want" (UNCSD 2012). Similarly, the International Union for Conservation of Nature (IUCN) identified "greening the world economy" as one of five key themes for its World Conservation Congress (IUCN 2012).

The extent to which the ideas and practices associated with neoliberalism have transformed conservation/development at all scales of activity has garnered considerable interest from critical scholars concerned with the tendency of capitalism to produce "accumulation by dispossession" (Harvey 2003, 2005).[1] Some of the most overt expressions of neoliberal capitalist expansion center on rural spaces in which private-sector interests move to control natural resources such as land, water, forests, minerals, and oil in order to develop new markets (e.g., Borras et al. 2011; Bridge 2008; Bakker 2007a; Sawyer 2004). Dispossession occurs, in part, when less powerful groups on the receiving end of capitalist expansion—such as small farmers and indigenous peoples—lose access to and control over the means of production.

Equally important, however, are the subtle and not-so-subtle ways in which neoliberalism has "colonized" conservation/development theory and practice over the past decade. The embrace of market-based approaches by a broad spectrum of conservation/development entities worldwide suggests that the assumptions and strategies associated with the green economy have coalesced into a largely unquestioned conventional wisdom. Thus, this chapter critically examines discursive manifestations of Nature™ Inc., pointing to the ways in which processes of neoliberalization within conservation/development arenas contribute to novel forms of governmentality (Foucault 2008; Fletcher 2010). My approach is to construct a conceptual interrogation of the term "capital" in order to explore how the logic of neoliberalism becomes enmeshed with and transforms the logic of conservation/development.

When viewed in toto, neoliberal conservation/development constitutes a performative arena within which markets are but one element alongside other social-structural factors that shape both targeted outcomes and power dynamics.[2] Neoliberal conservation/development manifests itself across scales of analysis within global-level environmental governance institutions such as the Convention on Biological Diversity (MacDonald and Corson 2012; MacDonald 2010a; Duffy 2006; Goldman 2005) and applied market-based instruments such as payment for ecosystem services and carbon offsets (McAfee 1999, 2012a; McAfee and Shapiro 2010; Bumpus and Liverman 2008) and within the context of specific programs related to protected areas (Brockington, Duffy, and Igoe 2008), ecotourism (Duffy

2002), and community-based conservation/development (Fletcher 2012a; Li 2007; West 2006), among others.

Amid the growing literature that critiques neoliberal conservation/ development, little attention has been directed at the ways in which the logic of capitalism has inscribed itself upon shifting theories and practices over time. In what follows, I examine the ways in which the term "capital" has been discursively extended and transformed as a means of articulating concepts and organizing practices related to sustainable livelihoods and institutional design/environmental governance. I position these two frameworks within the context of early discussions regarding sustainable development during the 1990s to uncover a progression of ideas that unfolded in conjunction with the rise of neoliberalism, thus facilitating the emergence of concepts and practices associated with the green economy.

For more than two decades, a range of actors focused on sustainable development—and, more recently, the green economy—have adopted the language of capitalism to project a vision that sees a mutually reinforcing relationship among economic growth, nature protection, and social equity objectives. The green economy seeks to value environmental goods and services in terms of natural capital, relying on a range of institutional "enabling conditions" to promote positive economic, ecological, and social outcomes. As with critiques of the initial framing of sustainable development in the 1987 Bruntland Report, deep skepticism lingers regarding the extent to which conservation/development approaches wedded to the logic of capitalism are internally contradictory and thus self-defeating (Fletcher 2012a; Buscher et al. 2012; Sullivan 2010a).

Beyond natural resources and services, however, the term "capital" became a ready metaphor to capture the range of "enabling conditions" that proponents of conservation/development might promote in order to achieve desired outcomes such as nature protection and poverty eradication. The so-called five capitals model (described below) offers a prime example of this tendency (see Porritt 2007). From a pragmatic perspective, the proposed integration of natural, social, human, manufactured, and financial forms of capital within conservation/development frameworks has garnered practitioners increased legitimacy and resources for activities aimed at stimulating local economies, empowering the rural poor, and protecting threatened ecosystems. From a critical perspective, however, reliance on constructs such as the five capitals raises questions about how frameworks centered on institutional design (environmental governance) and rural livelihoods (sustainable development) might hide key dimensions of the workings of capitalism and the skewed power relationships they often produce. By gaining greater intellectual purchase in both academic and international development deliberations, have such efforts disguised the actual work that capital performs?

I argue that the discursive extension of the term "capital" beyond its original meaning within economics tends to erase power in two ways. First, it hides the power dynamics associated with the exchange, flow, and accumulation of power resources. Second, it masks an incremental process of economic reductionism within conservation/development theory and practice. I construct my argument by revisiting the work of Pierre Bourdieu (1986), who offered one of the first presentations of different forms of capital. In building on Marx's ([1867] 1976) critique, Bourdieu both set a precedent for the conceptual expansion of the term "capital" and offered a nuanced analysis suggesting how capital constitutes different forms of power. By recovering Bourdieu's perspective on capital, I show that contemporary discursive practices that see accumulations of multiple forms of capital as a pathway to empowerment often ignore the everyday power dynamics and relationships that produce and reproduce social inequality and conflict over time.

The implications of critically examining the work that the term "capital" performs within conservation/development frameworks are far-reaching, since the construct has been largely subsumed within a broader discursive formation understood as "the green economy." Whereas early frameworks focused on livelihoods and institutional design deployed the word "capital" to connote "capacity" or empowerment, either within or independent of market exchanges, proposals for a green economy directly align conservation/development theory and practice with the logic of capitalism, suggesting the predominance of a deeply seated neoliberal environmentality (see Fletcher 2010). This combined paradigmatic shift and discursive slippage is important, because the conceptual extension of capital continues to connote natural, social, human, and cultural capacities even as the metanarrative of conservation/development theory and practice has shifted almost entirely to the logic of neoliberalism. Thus, despite a more explicit role for market-based approaches—and by extension the accumulation of economic capital—the continued use of the term "capital" to describe diverse human and natural capacities shields from view how economic capital accumulation unfolds.

The concept "capital"—as applied within sustainable livelihoods and institutional design frameworks—is understood mainly as any material or virtual asset or holding (a thing), as opposed to a continuing flow of goods and services tied to market exchanges (a process). By casting capital as largely stationary and independent from capitalist production, conservation/

development frameworks often mask many of the political-economic practices they seek to temper through empowerment. In contemplating the
role of different forms of capital in relation to social class and educational
achievement, Bourdieu (1986) used the terms "dissimulation" and "misrecognition" to capture this type of conceptual detachment where social
and cultural capital mask the power dynamics (flows, accumulations, class
relations) inherent to economic capital.

In this chapter, I uncover a paradox in Bourdieu's writing on "the forms
of capital" as a means of further illuminating the contradictions inherent
to conservation/development frameworks, including contemporary formulations of the green economy. Bourdieu elaborated multiple forms of
capital as a means of critiquing what he saw as the economization of social
inquiry, in which all social exchanges are seen as mercantile exchanges.
He sought to make visible a broader diversity of social interactions and
persistent inequalities by simultaneously expanding and critiquing how
the language of economics was deployed within the social sciences. The
paradox lies in the fact that by developing concepts like social and cultural
capital, Bourdieu contributed to the very intellectual developments he
sought to undermine. As I describe below, proponents of conservation/
development frameworks uncritically adopted and adapted different forms
of capital but for the most part did not account for the critique of capital
that Bourdieu and others emphasized. The concepts persisted even as the
political-economic context shifted but the critique was lost.

In order to recover the critique of capital relative to conservation/
development frameworks and the green economy, I first examine how different forms of capital emerged as the "building blocks" of conservation/
development during the 1990s and 2000s, drawing on two prominent frameworks: sustainable livelihoods and institutional design. In this context, capital constitutes an empowering asset that facilitates community development
and environmental governance, respectively. I contend that while the term
"capital" provided a convenient and uncontroversial shorthand for empowerment, its usage ignored the internal contradictions of capitalism and thus
elided the social and ecological ills the frameworks sought to redress.

The second part of the chapter returns to Bourdieu's writing on the
forms of capital in order to reevaluate his critique of economic reductionism and to suggest how his commentary on the workings of capital might
inform critiques of neoliberal conservation/development. A careful reading
of "The Forms of Capital" in relation to Bourdieu's wider work on practice
theory reveals the twin processes of dissimulation and misrecognition, in
which social actors intentionally and unintentionally disguise or euphemize

how everyday social interactions tend to reproduce persistent inequalities, particularly as they relate to the accumulation of economic capital.

The third section of the chapter explores how dissimulation and misrecognition of different forms of capital play out with respect to specific conservation/development interventions. Drawing on three examples from the literature, I highlight how the conflation of different forms of capital with empowerment in the context of neoliberal conservation/development programs hides the ways in which economic capital flows and accumulates. It is in this sense that dissimulation and misrecognition contribute to what I call the erasure of power.

The Building Blocks of Conservation/Development

In this section I unpack the concept of capital in relation to sustainable livelihoods and institutional design/environmental governance frameworks, each an important contributor to conservation/development thinking. I am mainly interested in formulations that present different forms of capital as building blocks or capacities that enable conservation/development. I compare and contrast three discussions of frameworks focused on sustainable rural livelihoods (Scoones 1998; Bebbington 1999) and institutional design (Brondizio, Ostrom, and Young 2009) that rely on multiple forms of capital. Each approach has strongly influenced conservation/development practice, although the former (livelihoods) incorporates capital centrally, while the latter (institutional design) uses the concept tangentially as synergistic with institutional design. While Scoones (2009) indicates that sustainable livelihoods approaches have declined in importance within international aid efforts, the five capitals remain deeply embedded in the conservation/development lexicon. Moreover, sustainable development frameworks that employ the five capitals model have extended beyond discussions related to international conservation/development, constituting the conceptual core of mainstream efforts to reform global capitalism by organizations like Forum for the Future (e.g., Porritt 2007).

In what follows, I critically examine how the three authors deploy the term "capital," how they understand the interrelationships among different forms of capital, and how they situate livelihoods and institutional design within broader political-economic arenas. It is important to note that I do not critique the frameworks per se but rather the role and function of different forms of capital within those frameworks. In certain respects, the authors recognize and discuss key aspects of my core argument. The underlying question is straightforward: Despite its usefulness in signaling the importance of ecological and social capacities, does the term "capital" hide more than it reveals?

Capital (Un)defined

Definition of the term "capital" is more elusive than it might seem. Formal definitions characterize capital as a "stock" or "reserve" of "accumulated goods" and distinguish it from income.[3] Thus, in most scholarly as well as popular usage, capital connotes an object; it is understood as a thing that can be accumulated and exchanged more or less freely. However, since the early nineteenth century, when the word's usage became linked to a particular economic system (ultimately providing the linguistic root of "capitalism"), a number of thinkers have exposed capital's enigmatic or illusory quali-ties.[4] One of Marx's ([1867] 1976) main contributions in this regard was to explicitly reconnect capital (as object) to the human labor that produced it, thus reflecting a particular mode of production (capitalism) and set of historically derived class relations. Ultimately, however, economists and noneconomists alike have stretched the original meaning of the term in an attempt to capture greater social and environmental complexity, but in doing so they have erased the relational and dynamic characteristics of capi-tal that Marx emphasized. The recent trajectory of capital as an explanatory concept within conservation/development studies illustrates this tendency.

The World Bank played a central role in defining and assessing sustain-able development in terms of different types of capital, attempting to place value not just on the exchange of money, goods, and services but also on nature, people, and social networks. For the World Bank (1997), concern over the "sustainability" of development emerged in response to critiques that conventional approaches focused solely on economic growth measured by the accumulation of material wealth like income or infrastructure. Initial adaptations accounted for the central role that training and education as well as environmental quality played in development by incorporating the terms "human" and "natural" capital.[5] Subsequently, World Bank staff also acknowledged the importance of social organizations and networks by add-ing another factor: "social" capital. In reports such as *Expanding the Measure of Wealth*, the World Bank (1997) proposed that sustainable development could be assessed across countries and regions based on relative endowments of these four types of capital: produced, human, natural, and social. In line with this approach, analysts understood capital to be an asset that could be accumulated, assigned a monetary value, and maintained or applied toward improving production and increasing wealth (see Bebbington 1999).

Beyond viewing the World Bank's efforts as a simple adaptation to critiques from civil society, it is important to briefly contextualize the organization's promotion of sustainable development and adoption of the capitals
framework in relation to the rise of neoliberal economic and political
policies globally. Ben Fine (2001) situates the emergence of "sustainable
development" and different forms of capital within the transition from the
Washington Consensus (orthodox neoliberalism) to the post–Washington
Consensus (reform neoliberalism). In the latter case, those responsible
for framing documents such as the Bruntland Report (WCED 1987), the
World Development Report 1990 (World Bank 1990), and *Expanding the
Measure of Wealth* (World Bank 1997) presented sustainable development
and, in some cases, multiple forms of capital to counter critiques regarding
environmental and social impacts. The use of terms like "social capital"
acknowledged the importance of social connectivity in empowering the
subjects of development but also turned attention away from the structural
inequities of neoliberal capitalism. Similarly, Goldman's (2005, 33) study
of the World Bank shows—among other things—how the organization
came to dominate knowledge production in conservation/development
arenas and thus contributed centrally to the formation of "regimes of green
neoliberalism" (see also Harriss 2002; Bebbington et al. 2004).

Assets and Capabilities

The World Bank's early attempts to create measures of natural, produced,
human, and social capital to supplement income as a measure of pros-
perity focused attention on capital as a material asset.[6] The sustainable
rural livelihoods framework expands on this approach, suggesting that the
presence or absence of "tangible and intangible assets" (forms of capital)
enables people to pursue different livelihood strategies (Scoones 1998, 7;
see figure 6.1).[7] Different manifestations of capital constitute a resource
base or endowment that drives productivity (use value). These assets may or
may not carry market value, although they may facilitate market exchanges.
I summarize six forms of capital and their applications in table 6.1.

In building on the work of Scoones (1998) and others, Bebbington (1999)
reframes the idea of capital somewhat to encompass its role both as a resource
that can enhance economic production and incomes and as capabilities
that may be instrumentally beneficial, meaningful, and emancipatory. This
construction emphasizes local agency and empowerment, where capabilities
potentially allow the subjects of development to "make a living, make living

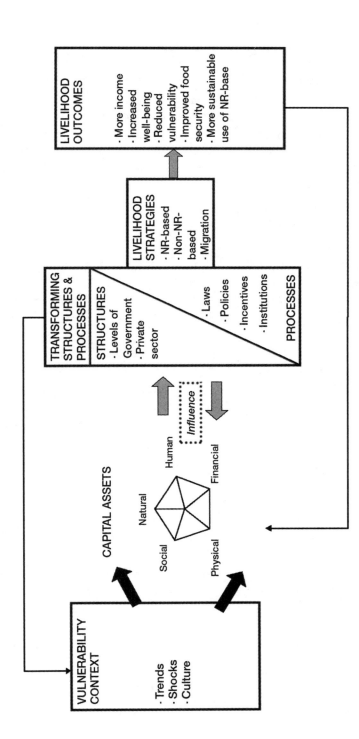

Figure 6.1. The Sustainable Rural Livelihoods Framework (redrawn by Leigh McDonald from Carney 1998; Farrington et al. 1999).

Table 6.1. Forms of capital and conservation/development frameworks

Form	Definition	Applications		Sources
		Sustainable livelihoods	*Institutional design*	
Natural capital	Stocks of natural resources (e.g., soil, water, air, trees) and environmental services (e.g., nutrient cycles, pollution sinks) that can be managed and enhanced for future gain	Soil fertility, forest reserves, grasslands, fisheries	Protected areas, carbon sinks (e.g., forests)	Costanza 2008; Natural Capital Project 2012; Millennium Ecosystem Assessment 2005
Physical capital	Human-made resources (e.g., roads, buildings, bridges, power plants) that can produce future wealth and well-being	Sawmills, offices, vehicles	Offices, vehicles	Brondizio, Ostrom, and Young 2009; Porritt 2007
Human capital	Knowledge and skills acquired through education and training that individuals apply to diverse activities	Capacity building	Capacity building	Brondizio, Ostrom, and Young 2009; Porritt 2007; Scoones 1998; Becker 1964
Economic capital	Monetary assets (e.g., cash, credit, stocks, bonds) that can be invested for future financial gain	Salaries, payments for ecosystem services, microcredit	Salaries, carbon trading	Porritt 2007; Scoones 1998
Social capital	Networks of trust and reciprocity that feature strong internal bonds within a social group and strong external bridges among social groups	Cooperatives, ecotourism enterprises, indigenous federations	Advocacy networks, NGOs, treaty organizations, certification bodies	Brondizio, Ostrom, and Young 2009; Bebbington 1999; Scoones 1998
Cultural capital	Durable norms and practices that shape identities, social interactions, and attachment to place	Subjectivities rooted in gender, race, ethnicity, and class	N/A	Bebbington 1999

meaningful, and challenge the structures under which one makes a living" (Bebbington 1999, 2000). As I discuss further below, I question whether use of the term "capital" as a metaphorical shorthand for "empowerment" actually works against people's ability to challenge structural inequalities.

Interestingly, Bebbington (1999) introduces an additional type of capital — cultural — that does not appear in mainstream discussions of sustainable rural livelihoods. The inclusion of cultural capital accounts for durable norms and practices that shape identities, social interactions, and attachment to place and might be important in explaining the motivations of indigenous federations or rural producer (campesino) cooperatives, among others. In noting that culture is not something that can be quantified, Bebbington argues that such qualities should be represented, since they often distinguish local from nonlocal understandings of poverty and development.

With regard to institutional design, Brondizio, Ostrom, and Young (2009) propose that different forms of capital enhance environmental governance across diverse organizational structures and scales, allowing actors to manage social-ecological systems more sustainably, effectively, and equitably. In this sense, individuals expend time and effort to build tools or acquire assets that can be applied toward increasing human welfare and ecological stability in the future. In drawing from the extensive literature on institutional design and common property (e.g., Ostrom 1990, 1994; Ostrom, Schroeder, and Wynne 1993; Ostrom, Gardner, and Walker 1994), Brondizio, Ostrom, and Young (2009) focus on how "human-made capital" (physical, human, social) enhances natural capital, as opposed to the sustainable livelihoods framework, which views natural capital in large part as a means to an end — sustainable development. They emphasize the importance of social capital in particular as providing the social "glue" that enables connectivity and resilience of governance regimes across multiple scales of organization.

Synergies, Trade-Offs, and Connectivity

Each of the authors contemplates how different forms of capital might interact and examines the possible outcomes that these interactions might produce. Scoones's (1998) presentation raises a series of questions regarding the sequencing of asset acquisition, the substitutability of types of capital, and the extent to which livelihoods strategies depend on multiple forms of capital working together (clustering). In extending this discussion, Bebbington (1999) recognizes the possibility of both synergies and trade-offs among different types of capital over space and time as people pursue livelihoods. For example, a positive synergy might occur where a producer organization allows an individual to complete training that enhances her income. A trade-off might occur when a family decides to decrease fallow cycles when planting crops or repeatedly seeks financial support from social networks without reciprocating in kind.

Along these lines, Bebbington (1999) and Brondizio, Ostrom, and Young (2009) emphasize the importance of social capital in enabling collective action. Such presentations highlight the structural qualities of social capital, linking it to other forms of capital as a means of capacity building. Related work by Bebbington and Perreault (1999, 395), for example, discusses social capital "in the form of community, federated, and national indigenous peoples' organizations and their institutional networks." In the context of highland Ecuador, social capital enhances community access to other forms of capital, thus presenting people and groups with greater development options. Similarly, Brondizio, Ostrom, and Young (2009) differentiate social capital from other forms of capital because it is hard to create and maintain social capital through external interventions, and it is very difficult to measure social capital in a meaningful way. In spite of these challenges, however, the authors argue that social capital can serve as a unifying concept that helps us understand institutional connectivity across multiscalar environmental governance regimes and interdependent ecosystems.

Despite these types of discussion that draw distinctions among the different types of capital, discursive extension of the term still creates an illusion of equivalency. This is particularly evident, for example, when practitioners reduce the more nuanced presentations of the sustainable livelihoods framework into guidance sheets, tool kits, and case studies (e.g., ELDIS 2011; IFAD 2011). Carney (2003, 27) observes that many applications "missed the point of SL [sustainable livelihoods] approaches" (poverty reduction through inclusive, people-centered development) by "going through the motions of using SL headings, reducing the holistic perspective to a set of rules that render the approach ineffective." The tendency is to assume that all capital assets are roughly equivalent and that access to and control of these assets—even though mediated by "transforming structures and processes" (figure 6.1)—will produce virtuous cycles of productivity (sustainable development). These presentations fail to recognize that the presence of such assets may actually contribute to the opposite outcome.

External Factors

Both the sustainable livelihoods and environmental governance frameworks situate the forms of capital in relation to "institutions and organizations"

(Scoones 1998), "transforming structures and processes" (Farrington et al. 1999; Carney 2003), and multiscalar social-ecological systems (Brondizio, Ostrom, and Young 2009). Initial formulations of the sustainable livelihoods approach referred to the ways in which organizations and institutions external to sustainable development sites mediate people's access to different forms of capital. Revised versions of the framework sought to unpack the black box of "transforming structures and processes" by distinguishing the roles of the private sector, government, international actors, and "wider society" (Carney 2003). Similarly, Bebbington (1999) situates access to and control of capital within a web of exchange relationships that is shaped by economic, social, and political "contingencies." In this sense, capital transactions unfold continuously across scales, while, at the same time, individuals' and groups' access to resources, other actors, and opportunities depends on an array of institutional constraints.

By contrast, in focusing on environmental governance, Brondizio, Ostrom, and Young (2009) place institutional design under the umbrella of social capital, viewing related challenges as a matter of developing institutional arrangements that effectively link different levels of organization. Their analysis of deforestation surrounding the Xingu Indigenous Park in northern Brazil suggests that external factors related to agropastoral development activities undermined the Xingu people's effective governance of the reserve. The authors present remarkable satellite imagery showing limited deforestation in the region surrounding the park in 1994 and extensive deforestation in the same area in 2005.

Most of the people involved with sustainable livelihoods and environmental governance frameworks are keenly aware of the importance of political-economic factors in shaping outcomes. Again, particularly in the case of the sustainable livelihoods approach, simplification of the capitals model ("asset pentagon" in figure 6.1) typically emerges when development organizations apply the framework in practice. Scoones (2009, 178) attributes oversimplification of the framework to the rapid rise of the sustainable livelihoods construct in 1998 from a "diagrammatic checklist" to a fully funded set of programs within the UK's Department for International Development (DfID's Livelihoods Department).

Since the sustainable livelihoods framework relies heavily on five forms of capital constituting an asset pentagon, it is important to note Scoones's (2009, 178) reflections on the model he helped to popularize, especially since they reinforce much of the analysis that I present in this chapter: "In some respects the focus on the 'asset pentagon' and the use of the 'capitals' metaphor was an unfortunate diversion. Other work on sustainable livelihoods emphasized other features . . . [such as] . . . the idea of institutions and organizations as mediating livelihood strategies and pathways. . . . They were subject to power and politics and were where the questions of rights, access and governance were centered." Although presentations of sustainable livelihoods and institutional design frameworks account for issues of access, trade-offs, and political-economic constraints, they do not contend with the underlying assumptions that inform the overarching project of development (but see Bebbington 2000). Thus, even if practitioners took the time to attend to political-economic forces, use of the term "capital" still erases structural manifestations of power that come into view in the work of Marxian and postdevelopment scholars (e.g., Smith 2008; Kovel 2002; Escobar 1995; Ferguson 1994). In other words, discursive extension of "capital" uncritically embeds empowerment frameworks within broader discursive formations associated with progress and modernity.

Moreover, to the extent that capital in any form is tied to development activities, those who rely on the term still must confront Marx's critique regarding the inherent contradictions of capitalism—particularly as they relate to primitive accumulation and environmental destruction (see Kovel 2002; Harvey 2010). These inherent contradictions point to the "fatal flaw" of capital extended. French social theorist Pierre Bourdieu contributed to both the expansion of capital as a concept and its critique by observing that social actors tend to hide the workings of power—and the conflicts and inequalities that they create—via a dual process that he presented in terms of "dissimulation" and "misrecognition."

Bourdieu's Extension of Capital

In this section I revisit Pierre Bourdieu's presentation of "the forms of capital" and contrast it with the formulations associated with sustainable livelihoods and environmental governance. Extension of the term "capital" represented a centrally important dimension of his theory of practice—a set of heuristics that he developed to analyze the production and reproduction of social inequality over time. I first summarize how Bourdieu defined three different forms of capital—economic, social, and cultural—and situated them within his theory of practice. I then discuss how he used the term "capital" as a surrogate for power. Finally, I examine how Bourdieu constructed the forms of capital as a critique of economic reductionism of social life. Each of these points offers insight into Bourdieu's use of the terms "dissimulation" and "misrecognition" to signify the ways in which individuals and groups engage in "collective denial" of the workings of capital.

Bourdieu (1986, 241) begins his exposition on "the forms of capital" with the following statement: "The social world is accumulated history, and if it is not to be reduced to a discontinuous series of instantaneous mechanical equilibria between agents who are treated as interchangeable particles, one must reintroduce into it the notion of capital and with it, accumulation and all its effects." Bourdieu's critical stance in these opening lines positions his work in the same sphere as Marx's writing on political economy. This is important because, in delineating multiple forms of capital, he sets out to show that economic capital can manifest itself in different ways but produce the same types of persistent inequalities associated with the workings of capitalism. In other words, Bourdieu maintained a conceptual link between the operation of economic capital within capitalist modes of production and the other forms he identified. The frameworks discussed in the previous section detach the accumulation of different types of capital from a particular mode of production.

Capital and the Workings of Power

Bourdieu's understanding of capital builds upon Marx's work by explicitly contemplating context, relationality, and power. It thus complements on one level but also deeply challenges more mainstream renderings linked to environmental governance and sustainable livelihoods. Bourdieu (1986, 241) defined capital as "accumulated labor." To the extent that actors appropriate capital, they capture "social energy" tied to productive effort over time and the social relationships that support it. He thus expanded upon Marx's critique of capitalism by describing multiple forms of capital that he associated with practices not typically categorized as economic.[8] Table 6.2 summarizes Bourdieu's definitions of the forms of capital and compares them to contemporary usages with sustainable livelihoods and environmental governance frameworks.

"The Forms of Capital" provides a discussion of social capital that highlights its dual character as both social structure and social process. Bourdieu (1986, 248–50) defines social capital as "the aggregate of the actual or potential resources which are linked to possession of a durable network of more or less institutionalized relationships of mutual acquaintance and recognition. . . . [It] provides each of its members with the backing of the collectively owned capital, a 'credential' which entitles them to credit, in the various senses of the word." Most references to Bourdieu in the social capital literature cite the first half of this definition and thus overlook the

Table 6.2. Comparison of forms of capital in Bourdieu's writing and conservation/development frameworks

Form	Definition	Comparison — Sustainable livelihoods	Comparison — Institutional design	Sources (by Bourdieu)
Natural capital	Given his focus on social relations, Bourdieu did not contemplate natural capital	Stocks of natural resources and environmental services		N/A
Physical capital	Bourdieu did not directly discuss physical capital but seems to include it as part of economic capital	Human-made resources that can produce future wealth and well-being	Not discussed	N/A
Human capital	Some aspects of Bourdieu's conceptualization of cultural capital would coincide with mainstream understandings of human capital; he critiques the idea of human capital as presented by Becker 1964	Knowledge and skills acquired through education and training that individuals apply to diverse activities		See Bourdieu 1986, 244
Economic capital	Bourdieu draws from Marx: "Capital is accumulated labor . . . which, when appropriated on a private . . . basis . . . enables agents and groups to appropriate social energy"	Monetary assets (e.g., cash, credit, stocks, bonds) that can be invested for future financial gain		Quote from Bourdieu 1986, 241
Social capital	Bourdieu: "Social capital is the sum of the resources, actual or virtual, that accrue to an individual or a group by virtue of possessing a durable network of more or less institutionalized relationships of mutual acquaintance and recognition"	Networks of trust and reciprocity that feature strong internal bonds within a social group and strong external bridges among social groups		Quote from Bourdieu and Wacquant 1992, 119; see also Bourdieu 1986, 248–49

Table 6.2. continued

| Form | Definition | ——— Comparison ——— | | Sources (by Bourdieu) |
		Sustainable livelihoods	Institutional design	
Cultural capital	Bourdieu: "Cultural capital can exist in three forms: in the embodied state, i.e., . . . long-lasting dispositions of mind and body; in the objectified state, in the form of cultural goods (pictures, books, . . . instruments, machines . . .), . . . ; and in the institutionalized state, . . . as will be seen in the case of educational qualifications"	Durable norms and practices that shape identities, social inter-actions, and attachment to place	Not discussed	Quote from Bourdieu 1986, 243; see also Bourdieu 1984, 1990

discussion that follows, which emphasizes social process. He notes that "these relationships may exist only in the practical state, in material and/or symbolic exchanges which help to maintain them; . . . [t]hey are more or less really enacted and so maintained and reinforced, in exchanges." Additionally, he emphasizes that "the reproduction of social capital presupposes an unceasing effort of sociability, a continuous series of exchanges in which recognition is endlessly affirmed and reaffirmed."

As I discuss below, the key characteristic of social capital in Bourdieu's rendering is not simply the presence or absence of a network or organization (collective capacity) but the ongoing enactment of exchanges among social actors who share a bond of trust (Wilshusen 2009). Different "species" of capital (Bourdieu's term) are akin to what Max Weber ([1922] 1978) called "power resources." Thus capital constitutes a dynamic flow rather than just a static stockpile (as the quotes from Marx and Harvey at the beginning of the chapter point out).

Despite adopting the language of economics in many instances (e.g., "collectively-owned capital," "credentials," and "credit"), Bourdieu did not frame exchanges as mere rational choices aimed at utility maximization. Rather, he understood human action to be embedded within culturally constructed "fields" or arenas of interaction. He used the term "habitus" to capture everyday practices tied to subjectivities (e.g., class) and life experiences. (For more on habitus, see Bourdieu and Wacquant 1992; Bourdieu 1990.) As a signifier of power, Bourdieu used the term "capital" in its various forms to portray individuals and groups as occupying positions of relative domination or subordination within specific fields.

While Bourdieu's presentation of social capital has received wide but superficial recognition, his writings on cultural capital have garnered much less attention. In addition to the power dynamics resulting from exchanges within networks, Bourdieu (1986, 243–48) maintained that another dimension of accumulation and social differentiation revolved around the relative distributions of cultural capital. He proposed that cultural capital could exist in three forms: an embodied state, an objectified state, and an institutionalized state. Embodied cultural capital refers to the incremental acquisition of capabilities and dispositions of mind and body. According to Bourdieu, the attainment of capacities (such as education) assumes long-term self-investment, although the process may be more a result of everyday practices as opposed to "deliberate inculcation." This type of cultural capital approximates what current frameworks call human capital; however, Bourdieu notes that attainment of embodied cultural capital depends on social-structural factors such as family and class (more on this below).

Objectified cultural capital comprises material and symbolic artifacts such as writings, paintings, machines, and instruments. In this regard, Bourdieu's usage coincides to a certain extent with contemporary under-standings of physical capital. At the same time, however, Bourdieu's (1986, 247) formulation emphasizes the relational character of capital in all of its forms, which highlights the power dynamics associated with what many think of as human and physical capital: "Cultural goods can be appropriated both materially—which presupposes economic capital—and symbolically—which presupposes cultural capital. It follows that the owner of the means of production must find a way of appropriating either the embodied capital which is the precondition of specific appropriation or the services of the holders of this capital." Similarly, institutionalized cultural capital connotes formally conferred status such as titles that are tightly intertwined with embodied and objectified cultural capital. For example, an individual might acquire an education (embodied) that provides access to economic capital more readily via academic credentials (institutionalized) represented by a diploma (objectified). Taken together, Bourdieu's description of three types of cultural capital is both more specific and far-reaching than Bebbington's (1999) usage. Whereas Bebbington emphasizes how cultural practices that ascribe meaning to identities and place enable and empower collective action, Bourdieu focuses attention on the ways that specific manifestations of culture contribute to distinction and differentiation.

Beyond relationality and flow, Bourdieu's conceptualization of capital is deeply contextual. He used the term "field" in addition to "habitus" to delineate arenas of struggle in which social actors seek to accrue economic, social, and cultural capital. He defined fields loosely to capture domains of social interaction such as the "artistic field" and the "academic field."[9] He linked manifestations of capital—such as the example of educational credentials offered above—to a particular field. This is important because it suggests that, beyond the competitive struggles of social actors, the power derived from capital depends upon the discursive and institutional practices associated with fields (i.e., structural power).

Given the multiple dimensions of power encapsulated in Bourdieu's use of the term "capital," one might ask why he chose to adopt the term within his theory of practice. The answer centers on his concern regarding what he saw as a take-over of social inquiry by a narrowly defined, "classical" understanding of economics. As I discuss below, Bourdieu's stance anticipated subsequent critiques regarding the neoliberalization of social theory (e.g., Fine 2001) and conservation/development frameworks.

Critique of Economic Reductionism

Bourdieu's differentiation of multiple forms of capital in part constitutes a critique of dominant, economistic understandings of social life (exemplified in the previous section by the World Bank framework). He sought to undermine the tendency of equating all social exchanges with mercantile exchanges by developing a counterterminology under the umbrella of a "general science of the economy of practices" (e.g., Bourdieu 1984). As I have noted, the examination of social processes that produce and reproduce structures of inequality was central to this project.

Bourdieu (1986, 242) viewed the economization of social inquiry as a reflection of hegemonic power structures: "It is remarkable that the practices and assets thus salvaged from the 'icy water of egotistical calculation' (and from science) are the virtual monopoly of the dominant class—as if economism had been able to reduce everything to economics only because the reduction on which that discipline is based protects from sacrilegious reduction everything which needs to be protected." Some of Bourdieu's rationale for employing the language of economics in his critique of economic reductionism emerges in his discussion of cultural capital. He notes, for example, that the notion of cultural capital emerged in relation to comparative research on scholastic achievement of children from different social classes. Contrary to dominant explanations from the 1960s and 1970s, which emphasized natural aptitudes and human capital (following Becker 1964), Bourdieu pointed to social-structural factors such as family background, class status, and institutional access as significant contributors to academic and employment success rates. He critiqued analyses that focused exclusively on monetary investments and profits and ignored other types of inputs and gains. He introduced the term "cultural capital" to both expand the scope of causal factors and enhance how scholars defined success.

In critiquing Becker (1964) and others, Bourdieu (1986, 244) concludes that "because they neglect to relate scholastic achievement strategies to the whole set of educational strategies and to the system of reproduction strategies, they inevitably, by a necessary paradox, let slip the best hidden and socially most determinant educational investment, namely, the domestic transmission of cultural capital." Thus, despite critiques that have found his writing to be overly deterministic and economistic, Bourdieu sought to contest purely behavioralist theories within the social sciences that tended to deposit power in the hands of autonomous individuals and groups while ignoring the broader social-structural constraints shaping choices. In this sense, he anticipated later critiques of the colonization of social theory by neoclassical economics (Fine 2001) as well as the neoliberalization of conservation/development (summarized in Castree 2008a, 2008b; Wilshusen 2010). Fine's (2001) interrogation of the rise of social capital, for example, argues that most applications of the term—such as those sponsored by the World Bank—emphasize a community's or a region's lack of collective capacity and thus target these places as being in need of interventions that would enhance their assets. Similar to Bourdieu's analysis of scholastic achievement, Fine argues that such approaches ignore wider political and economic factors that structure access to resources and long-standing inequities.

Ultimately, Bourdieu's use of the term "capital" pivots on the notion that social life is far more complex than narrow interpretations of *Homo economicus* allow. In relation to extending the definition of capital, he saw that social actors tended to compete for resources that were rooted in economic capital but were not necessarily convertible into money (e.g., social and cultural capital). He observed also that people tended to disguise or "euphemize" everyday power relationships. This raises the question: What types of social interactions and outcomes does the discursive extension of capital produce?

Dissimulation and Misrecognition

Taken as a whole, Bourdieu's theory of practice represents a far-reaching alternative to economistic understandings of social life. By choosing the term "capital" to capture both power dynamics and historically rooted social relationships, he deliberately used the language of economics to make visible what mainstream economics (defined in terms of capitalism) conceptually erased. Bourdieu (1986, 242) explains:

> Economic theory has allowed to be foisted upon it a definition of the economy of practices which is the historical invention of capitalism; and by reducing the universe of exchanges to mercantile exchange, which is objectively and subjectively oriented toward the maximization of profit, i.e., (economically) *self-interested*, it has implicitly defined the other forms of exchange as noneconomic, and therefore *disinterested*. In particular, it defines as disinterested those forms of exchange which ensure the *transubstantiation* whereby the most material types of capital—those which are economic in the restricted sense—can present themselves in the immaterial form of cultural capital or social capital and vice versa.

In other writing, Bourdieu (1977, 1984, 1990) ties this illusory transmutation of economic capital into social or cultural capital as a discursive practice, where representations of social and cultural capital take on an inherent legitimacy or prestige and thus mask their roots in economic capital. Actors' "misrecognition" or "collective denial" of the transubstantiation of different types of capital in their symbolic form constitutes a process that Bourdieu understood as the enactment of symbolic power. In this sense, actors participate in the unstated but mutually understood theatrics of denial, which in turn legitimize social practices and power relationships in a given field of play.

Bourdieu's (1977, 195) analysis of gift exchanges, which draws from his ethnographic fieldwork among the Kabyle of Algeria during the late 1950s, provides some context for both the critique of economistic social science and the observation regarding transubstantiation of different forms of capital: "Wealth, the ultimate basis of power, can exert power, and exert it durably, only in the form of symbolic capital; in other words, economic capital can be accumulated only in the form of symbolic capital, the unrecognizable, and hence socially recognizable, form of the other kinds of capital."[10] While the theory of practice offers a deeply nuanced view of social process and power relationships, Bourdieu's choice to counter narrow, economistic views of social life by using terminology derived from economics (fighting fire with fire) has contributed centrally to capital's transformation into a paradox, a type of misrecognition in which complex interactions are misrepresented as assets of empowerment. Bourdieu (1986, 243) makes clear that he intends the term "capital" to be a heuristic device to allow those engaged

148 • Peter R. Wilshusen

in critical social inquiry to see past the illusory yet hegemonic conflation of "mercantile exchanges" with the entirety of social exchanges: "A general science of the economy of practices, capable of reappropriating the totality of the practices which, although objectively economic, are not and cannot be socially recognized as economic, and which can be performed only at the cost of a whole labor of dissimulation or, more precisely, *euphemization*, must endeavor to grasp capital and profit in all their forms and to establish the laws whereby the different types of capital (or power, which amounts to the same thing) change into one another."

The "dissimulation" and "euphemization" of cultural or social capital to cover for economic capital produce the "social alchemy" (collective denial or misrecognition) that forms the core of Bourdieu's dramaturgical understanding of social life.[11] The great irony is that, by constructing a heuristic—capital—to capture power dynamics and class relationships, Bourdieu set the stage for other scholars, representing schools of thought rooted in classical economics, to engage in exactly the acts of dissimulation that he sought to contest. As a result, many confer magical properties upon different forms of capital, using the term to represent power, ipso facto, rather than as a construct that captures flows of power resources and configurations of power relationships. The impact of discursively extending capital beyond its roots in monetary exchange is the erasure of power. In other words, the conceptual detachment of the different forms of capital from the workings of economic (or financial) capital depoliticizes conservation/development interventions, whether framed in terms of sustainable livelihoods, institutional design, the green economy, or something else.

The Social Alchemy of Capital

With Bourdieu's critique of economic reductionism and insights on the workings of capital in view, the question then becomes: How do dissimulation, misrecognition, and the erasure of power occur in practice? Practice theory builds on the literature in critical development studies by highlighting the production and reproduction of power relationships over time. (In addition to Bourdieu's work, see Ortner 2006; Sewell 2005; and Mosse 2010.) In critically examining the discursive extension of capital within conservation/development frameworks, I turn attention in this final section to the ways in which the erasure of power unfolds within the broader context of a neoliberal political economy. How does the logic of neoliberal capitalism manifest itself both symbolically and materially in specific contexts? How do the subjects of conservation/development activities coproduce narratives of empowerment and nature protection alongside those positioned as change agents (e.g., the state, private-sector firms, NGOs, aid organizations)? How do dissimulation and misrecognition shape social relationships and impact conservation/development designs?

The neoliberalization of conservation/development over the past decade in particular has produced a significant realignment of ideas, actors, and institutional configurations as well as a concomitant set of responses by communities, NGOs, and state agencies to tie their activities to the logic of the market. Thus, on one level, "the market" and related flows of economic capital appear more overtly in programs like payment for ecosystem services (PES) and reduction of emissions from deforestation and forest degradation (REDD+). On another level, however, the growing literature on neoliberal conservation/development uncovers how market-based approaches disguise their core assumptions and hide power dynamics in ways that work against stated goals such as poverty eradication and nature protection. In this sense, the literature mirrors Bourdieu's critique of economic reductionism and his observations on the workings of capital. In responding to the questions posed above, I draw upon three recently published pieces to suggest how the erasure of power plays out in practice. As an initial comparison, table 6.2 juxtaposes Bourdieu's definitions of different forms of capital alongside the ways that capital appears in sustainable livelihoods and institutional design framework—most of which carries over to the green economy.

The Vagaries of Social Capital

My work (Wilshusen 2009) on the everyday politics of community forestry in Quintana Roo, Mexico, uses Bourdieu's theory of practice explicitly and thus incorporates his perspectives on the forms of capital generally and social capital in particular. Social capital offers perhaps the most prominent example of dissimulation and misrecognition within the context of conservation/development activities. The term attained near-universal recognition and usage during the 1990s and continues to be a key concept for conservation/development practitioners. In studying the responses of rural land-grant communities (*ejidos*) in southeastern Mexico to neoliberal policy reform measures during the late 1990s and early 2000s, I found that both state and nonstate actors relied on the idea of social capital to organize activities aimed at encouraging community members to expand and strengthen their market-based activities.

One example of such an attempt at social capital "formation" in the context of a regional community forestry association illustrates how dissimulation and misrecognition occur in practice. Between 1996 and 2001 a Mexican federal agency channeled approximately US$265,000 in grants to support a timber-marketing fund that would allow communities to aggregate their wood products, enhance their collective bargaining power, and pursue higher selling prices in the marketplace. Representatives from each of the main actor types—federal agency, forestry association, communities, and state government—portrayed the timber-marketing fund as an example of an intervention that empowered rural producers to strengthen their capacity as individual entrepreneurs and members of collective enterprises. By 2002 the fund had received national-level attention within policy circles and was included as a successful case study in an Interamerican Development Bank (IADB) study on enterprise development.

As a participant-observer working with the community forestry association mentioned above for most of 2000, I analyzed the fund's accounts and found that the collective benefits recounted by promoters were far outweighed by a considerable number of unsettled transactions with a wide range of actors. Specifically, 92 percent of the $265,000 that the fund received during a five-year period circulated as informal loans to individuals, cash advances to communities, and open accounts with timber wholesalers. The majority of the funds (63 percent) were never recovered, leaving the timber-marketing fund insolvent by late 2001 (Wilshusen 2009). Thus, even as the fund was crumbling under the weight of unpaid loans and other unresolved transactions, many of the principal actors involved were still presenting it as a success story to interested parties such as the IADB.

As I note elsewhere (Wilshusen 2009), social capital played an important role in how the fund operated day-to-day—but not in the way that the fund's promoters reported to external audiences. The narrative offered to external audiences emphasized the fund's role in empowering rural producers to more effectively interface with market actors, much in the way that social capital is typically discussed in conservation/development circles. Behind the scenes, however, the fund supported an active network of formal and informal loans and other transactions that produced significant inequalities in terms of economic capital accumulation. The fund generated the social alchemy or "collective denial" that Bourdieu described in that most observers saw it only as social capacity while ignoring the actual flow of resources (timber and money) and related power shifts.

Within the context of community forestry in southeastern Mexico, the logic of neoliberalism intersected in complex ways with long-established practices built from the cloth of collectivism (see Wilshusen 2010). Social capital formation emerged in the late 1990s as a means of "capacity building" within and among rural communities that dovetailed with the neoliberal turn in Mexico's agrarian sector. In addition to state and NGO actors that promoted social capital formation, many rural producers embraced the idea of becoming more successful entrepreneurs while simultaneously eschewing most aspects of collective forest management, which had been the dominant approach. The example of the timber-marketing fund suggests how the concept of social capital assists with the neoliberal project of building entrepreneurial capacity at the same time that it disguises the workings of economic capital.

Conjuring Natural Capital

Within the scope of the green economy, natural capital takes center stage. The term "natural capital" redefines broader constructs of "nature," "environment," and "biodiversity" in terms of its component goods, such as food, fuel, and minerals, and its ecological services, such as water regulation and carbon storage (see Porritt 2007; Natural Capital Project 2012). UNEP's (2011) *Towards a Green Economy* report, for example, emphasizes that "a green economy recognizes the value of, and invests in, natural capital" as one of its main findings. The assumption that greater legibility or internalization of environmental goods and services in pricing will lead to nature protection and, in some presentations, poverty reduction is widely accepted in conservation/development circles but largely untested. In particular, Bourdieu's notion of "transubstantiation," discussed above, allows us to ask how the transformation of nature into natural capital hides the flow and accumulation of economic capital but also reconfigures conservation/development theory and practices into a predominantly neoliberal exercise—what Fletcher (2010) calls neoliberal environmentality.

MacDonald and Corson's (2012, 159) critical examination of The Economics of Ecosystems and Biodiversity (TEEB) project explains the process of transubstantiation in terms of "virtualism," concluding that measuring and valuing natural capital "makes nature legible by abstracting it from social and ecological contexts and making it subject to, and productive of, new market devices." Using the 10th Conference of the Parties (COP10) to the Convention on Biological Diversity (CBD) as a stage, the authors show how a range of key actors perform and socially construct green economy frameworks such as TEEB, simultaneously merging virtual and real nature within the concept of natural capital and disguising its role as economic capital (dissimulation and misrecognition in Bourdieu's terms).

MacDonald and Corson's analysis points to the ways in which the realization of an idealized model of a green economy based upon natural capital is predicated on new methods of privatizing access to and control of nature and the invention of new nature-based commodities and markets. TEEB represents a prominent example of a wider trend toward aligning conservation/development theory and practice with the logic of neoliberalism. The project began in 2007 as a discrete study of the economics of biodiversity loss but rapidly evolved by 2010 into a broader policy instrument formally tied to the CBD's governing secretariat. The core assumptions of neoliberal capitalism are expressed more explicitly in this case compared to the example of social capital formation discussed above. Projects like TEEB provide a tidy framework that allows policymakers, practitioners, and others to first measure and then assign monetary value to nature's components and processes. The problem of biodiversity loss from this perspective is understood as a matter of appropriately pricing nature (natural capital) within a given market.

While MacDonald and Corson's global-scale analysis of natural capital and TEEB differs from my local-scale examination of social capital relative to community forestry in southeastern Mexico, both studies turn attention to the performative aspects of the social alchemy of capital and the shifting power dynamics that these performances produce. Similar to the public presentations of the timber-marketing fund that emphasized its role in empowering rural producers to engage in market transactions, proponents of TEEB actively sought to legitimate both the project and the process of monetarily valuing natural capital by suggesting how pricing enabled market-based approaches such as payment for ecosystem services. By quantifying nature's value, advocates strategically positioned conservation as a benefit rather than a cost vis-à-vis development. Moreover, TEEB's main supporters argued that since nature is the "GDP of the poor," its efforts to account for nature's value in numerical terms would thus serve the interests of the poor (MacDonald and Corson 2012, 176). Hidden from view in this presentation are the myriad examples of accumulation by dispossession that emerge in the wake of privatization, commodification, and marketization.

Although Bourdieu did not contemplate natural capital in his writing, the process of transubstantiation or transmutation that he identified with respect to social and cultural forms of capital mirrors MacDonald and Corson's understanding of natural capital as an expression of virtualism. By constructing an abstraction—natural capital—as a surrogate for both nature and economic capital, proponents of the green economy have incrementally and almost completely realigned conservation/development theory and practice with the core dictates of neoliberal capitalism. As key examples of what Bourdieu called symbolic capital, terms like "the green economy" and "natural capital" take on inherent legitimacy via multi-faceted collective performances such as the rollout of TEEB at COP10. Natural capital's shape-shifting characteristics (active dissimulation in this case) leads to misrecognition of its roots in economic capital and the legit-imation of a whole new sphere of ideas, actors, and institutions organized in terms of neoliberalism.

Conservation/Development qua Neoliberalism

The process of neoliberalization that MacDonald and Corson (2012) identify with respect to TEEB and the proceedings of the Convention on Biological Diversity mirrors Bourdieu's critique of economic determin-ism in social inquiry discussed above. Bourdieu's concern that all social exchanges were understood as mercantile exchanges led him to elaborate a more general "economy of practices" that included multiple forms of capital. A similar concern emerges in critiques of neoliberal conservation/development in which a widely shared conventional wisdom has emerged that embraces the market-based approaches of the green economy.

In this regard, Fletcher and Breitling's (2012) review of two market-based approaches in Costa Rica's Osa Peninsula—ecotourism and PES—argues that neoliberal strategies create an oppositional tension between conserva-tion and development objectives rather than the mutually reinforcing rela-tionship that advocates promote. In asking whether the "tools" of capitalism can conceivably resolve the environmental and social problems that their application tends to perpetuate, Fletcher and Breitling posit that neoliberal conservation/development techniques function as an "evasion of inequality." While his analysis does not center directly on the discursive extension of capital, it does suggest how narrowly economistic attempts at integrating conservation and development produce collective denial of social inequality.

While nature protection efforts focused on the Osa Peninsula have succeeded in reducing forest cover loss and fragmentation measured since the late 1970s, it is unclear whether or not ecotourism and PES have contributed to this trend. For example, ecotourism in the Osa may contribute to increased local income, enhanced conservation conscious-ness, and positive attitudes toward conservation activities. At the same time, however, legal restrictions appear to influence declines in deforestation more directly, while the benefits of ecotourism enterprises and landownership have accrued to foreign entrepreneurs. Similarly, studies of the Osa's PES projects, which make direct payments to landowners for protecting forests, suggest that reductions in deforestation have more to do with preexisting legal restrictions on forest clearing rather than economic incentives produced by payouts. Like ecotourism, PES appears to disproportionately benefit wealthier landowners. Moreover, PES activities in the Osa continue to depend heavily on injections of funds from the Costa Rican government, the World Bank, and the Global Environment Facility (GEF), raising questions as to whether they constitute a self-sustaining, market-based approach (Fletcher and Breitling 2012).

Ultimately, Fletcher and Breitling (2012) conclude that neoliberal conservation measures on Costa Rica's Osa Peninsula evade the issue of inequality by ignoring the unbalanced accumulation of economic capital cited above and turning attention away from large-scale industrial activities that pose the greatest threat to the Osa's forests. Thus the market-based designs represented by ecotourism and PES constitute a type of euphemism or dissimulation that promises increased incomes, entrepreneurial training, and stronger support networks (economic, human/cultural capital, and social capital, respectively) that will both enhance economic development and protect nature. As Fletcher and Breitling expose, however, this dissimulation turns out to be illusory on multiple levels: the programs do not produce conservation or development, they may not be market-based mechanisms (in the PES example), and they ignore the root causes of ecological and social degradation.

Whether it emerges more explicitly (e.g., the five capitals framework) or more implicitly (e.g., UNEP's *Towards a Green Economy*), the discursive extension of capital plays a central role in the neoliberalization of conservation/development, helping to produce a "commonsense" narrative about empowerment and nature protection that goes largely unchallenged. The language of capital subtly reinforces the assumptions of neoliberal capitalism without saying its name. As the examples in this section illustrate, while the logic of neoliberalism currently dominates conservation/ development thinking, it is not monolithic — it unfolds in complex, differentiated ways across a range of fields or arenas ranging from global meetings such as the COP10 to the CBD to community-level interventions in places such as rural Mexico. The tendency of those involved with these negotiations and interventions to first dissimulate and then misrecognize the workings of capital in its different forms hides the internal contradictions of neoliberal capitalism and masks the dissent that often emerges in response to the material manifestations of these contradictions—environmental decline and social inequality. This is what I have called the erasure of power.

Conclusion

Bourdieu composed his chapter on the forms of capital prior to the neoliberal turn, but his critique of economic determinism and the dissimulation and misrecognition of capital in the context of everyday social interactions offers important insights with respect to neoliberal conservation/development. Indeed, in a series of essays published in the late 1990s and early 2000s, Bourdieu (1998a, 2003) offered a scathing critique of neoliberalism and what he considered to be "the tyranny of the market."

While the five capitals model may have already lost some of its cachet in discussions of sustainable livelihoods (Scoones 2009) and plays a tangential role in institutional design frameworks focused on common property and environmental governance, the construct remains strongly embedded both explicitly and implicitly within conservation/development discourse and practice. UNEP's (2011) *Towards a Green Economy* targets adequate pricing of natural capital but further emphasizes the fundamental importance of enabling conditions defined in terms of capacity building, education, training, strengthened governance, and sound regulatory frameworks, among other things. This approach builds directly on previous formulations of "natural capitalism" (e.g., Porritt 2007) that place the five capitals at the center of a framework for sustainable development.

In this light, advocates of the green economy or the five capitals model may view my critique as overwrought. From such a "green" perspective, one might argue that capitalism can manifest itself in a number of ways; it does not necessarily have to produce persistent inequality and environmental destruction if those involved pursue "good" growth (e.g., health and happiness) as opposed to "bad." Similarly, one could claim that use of the five capitals construct within sustainable livelihoods and institutional design frameworks has helped to improve upon traditional, top-down development approaches and channeled scarce resources toward the ends of empowerment, nature protection, and good governance.

Yet, as the conceptual discussion and summary examples in this chapter show, discursive extension of the term "capital" produces an illusory projection of empowerment, social unity, and ecological resilience that ultimately feeds a form of economic determinism—neoliberal conservation/

development. By construing capital as a thing rather than a process, it erases the workings of power in practice, hiding the internal contradictions of neoliberal capitalism at the same time (see McAfee 2012a). From this point of view, while natural capitalism may be kinder and gentler than extractive and exploitative manifestations, it still leads to dissimulation and misrecognition of the true face of capital. In this sense, even though Bourdieu's work helped set the stage for the very problem he sought to preempt, it also provides a lens that captures how capital flows and accumulates in the context of market and other types of exchanges. The paradox that I identify in Bourdieu's writing on the forms of capital plays out similarly with respect to the green economy. The extent to which conservation/ development theory and practice have taken on the character of neoliberal capitalism has entailed a wholesale adoption of the logic and discourse of the market—the building blocks of Nature™ Inc. An untenable paradox emerges at the point when proponents of neoliberal conservation/ development gain the legitimation that grants access to centers of power but cannot achieve nature protection or poverty alleviation.

Notes

Earlier versions of this chapter were presented at the April 2011 meeting of the Association of American Geographers (AAG) held in Seattle, Washington, and at the June 2011 conference "Nature™ Inc.: Questioning the Market Panacea in Environmental Policy and Conservation" held at the International Institute of Social Studies (ISS), The Hague, Netherlands. This chapter draws from work that was originally financed by the Social Science Research Council (SSRC), with funds provided by the Andrew W. Mellon Foundation, a Fulbright–Garcia Robles grant, an Inter-American Foundation fellowship, and travel grants from Bucknell University. The David and Patricia Ekedahl Professorship in Environmental Studies at Bucknell University also provided support for writing and travel.

1. I use the term "conservation/development" to capture a range of practices and approaches that explicitly link nature protection and sustainable development objectives and carry names such as integrated conservation and development projects (ICDPs), community-based conservation (CBC), and community-based natural resources management (CBNRM). In many cases these activities employ market-oriented strategies associated with neoliberalism.

2. Fletcher (2012c, 297) describes five characteristics of neoliberal conservation/ development, including (1) devolution of governance responsibilities from state actors to nonstate actors such as NGOs, (2) creation of markets centered on natural resources, (3) privatization of natural resources, (4) commodification of natural resources, and (5) partnerships with private corporations for fund-raising purposes.

3. The *Oxford English Dictionary* (OED) offers the following entry on capital: "The accumulated wealth of an individual, company, or community, used as a fund for carrying on fresh production; wealth in any form used in producing more wealth." The *OED* traces the word's monetary usage to a 1611 entry in Randle Cotgrove's *A Dictionarie of the French and English Tongues*: "wealth, worth; a stocke, a man's principall, or chiefe, substance."

4. See Raymond Williams's ([1976] 1983, 50–52) entry on "capitalism."

5. Although the World Bank grouped the different types of capital under its sustainable development framework, it did not invent the terms. Becker (1964) popularized "human capital," especially within economics. "Social capital" emerged in the sociological literature in the 1970s and 1980s (e.g., Coleman 1988; Bourdieu 1986) before achieving widespread visibility in the 1990s, especially following Putnam (1993). The term "natural capital" is most closely associated with the subfield of ecological economics (e.g., Jansson et al. 1994).

6. The designation "produced" capital used by the World Bank combines "financial" capital and "physical" capital (e.g., infrastructure)—a distinction that appears in subsequent presentations of the five capitals.

7. For historical context on the sustainable livelihoods approach, see Scoones (2009).

8. My summarization of Bourdieu's conceptualization of capital is not intended to be comprehensive. Rather, I seek to capture certain distinctive qualities that emerge from his usage of the term, especially in his chapter "The Forms of Capital" (1986).

9. Further discussion of Bourdieu's formulation of fields can be found in Bourdieu and Wacquant (1992).

10. Bourdieu's use of the term "symbolic capital" refers to practices of representation that confer enduring and, in many cases, unquestioned legitimacy upon the other forms of capital. It bears a resemblance to Gramsci's (1971) use of "common sense" in his writing on hegemony to the extent that the interests of capital become the interests of all.

11. The term "social alchemy" appears in Bourdieu (1977, 195). The influence of Marx's formulation of "commodity fetishism" seems clear in Bourdieu's construction inasmuch as Marx deliberately used the term "fetish" to signal the illusory or magical transformation achieved by commodities when they take on inherent value, thus masking the social relations and labor that produced them. It is interesting, therefore, to note Ben Fine's (2001, 16; 2010) ongoing critique of social capital, which refers to the concept as a form of fetishism.

CHAPTER SEVEN

Performative Equations and Neoliberal Commodification

The Case of Climate

Larry Lohmann

The Dilemmas of Theory

Between the insight that current economic and environmental crises are being exacerbated by the new forms of commodification characteristic of neoliberalism and the detailed specification of what those forms are lies the work of a hundred lifetimes. Commodification is a many-splendored process, and it has to be. All commodities-in-the-making are different, and so are the series of acts and actors, impulses and resistances that contribute to, or block, their making or unmaking. The proliferation of ambitious, variously contested commodities that has sprung up in the neoliberal era—from wetland offsets (Robertson 2000, 2004) and collateralized debt obligations to genome information products (Sunder Rajan 2006), public services (Huws 2011), and species (Pawliczek and Sullivan 2011)—only amplifies the diversity. As the work of scholars as varied as Elinor Ostrom (Ostrom, Gardner, and Walker 1994), Viviana Zelizer (1995), Colin Williams (2005), Margaret Radin (1996), and Brett Frischmann and Mark Lemley (2006) confirms, the idea that there exists a single, uniform process of commodification operating everywhere on the as-yet uncommodified is as unfounded as the quasi-deistic notion, equally emblematic of the neoliberal era, that everything already is a commodity (O'Connor 1994a).

The shorthand "the commodification of nature" is loaded with a particularly great breadth of meanings. If deployed without awareness of the teeming multitude of differing cases, each with its own complexities, the term runs the risk of confusing and clarifying in equal measure. Karl Polanyi (1944) and John Maynard Keynes (1936), following a path opened by Marx, highlighted some of the distinctive features and pitfalls of the commodification of land. In the neoliberal era, Marxist-inspired thinkers, actor-network theorists, and others have revealed some of the diversity of the "black boxes" that have to be opened to expose the predicaments specific to the commodification of many other aspects of "nature" (e.g., Kloppenburg 1988; Bridge 2000; Holm 2001; Boyd 2001; Martinez-Alier 2003; Henderson 2003; Robertson 2004; Mansfield 2004; Bakker 2004; Robbins and Luginbuhl 2005; O'Neill 2006) — including those occasioned by various types of what Martin O'Connor (1994a) calls "nature's resistance" or what Noel Castree (2003, 285) calls "contradictions between the materialities of nature and those of the commodification process." Continually reminding such scholars of the particularities of individual struggles over commodification is a worldwide spectrum of confrontations at the grass roots over issues ranging from the enclosure of community forests to the expansion of credit involved in microfinance.

Nowhere is attentiveness to the diversity of commodification more crucial than in the formulation of environmentalist strategy. "Our Earth is not for sale" may be a good rousing slogan for Friends of the Earth International, "tu no puedes comprar el sol" a felicitous line in a popular anticapitalist anthem by the Puerto Rican group Calle 13, and "Nature™ Inc." an excellent title for an international conference of critical academics on problems connected with current trends in the capitalization of nature. But without extensive explication, such throwaway phrases are too abstract to give much idea of where to locate the challenges and opportunities that are exercising so many movements and thinkers today or of where and how to make critical interventions. In reality, Nature (whatever one might mean by this questionable "key word") has been Incorporated in one form or another for a good long time, and various bits of Earth have been on the block for many centuries. What, if anything, is really new, and if what is new is as frightening as often claimed, what is to be done about it?

Formal definitions of commodification are of limited help: their concision tends to be in inverse proportion to their applicability. Take, for example, the definition offered by Karen Bakker, one of the subtlest scholars of the commodification of water. She is at pains to dispel what she rightly regards as confusions among commodification, privatization, and commercialization. Commodification, Bakker says, is the "creation of an economic good, through the application of mechanisms to *appropriate* and *standardize* a class of goods and services, enabling them to be sold at a price determined through market exchange" (2007b, 103, emphasis added). One of the virtues of this definition is that it pinpoints the enduring prominence, in commodification, of ownership, control, and measurement. Yet such definitions are considerably less illuminating today, in an era of financialization and a growing "green economy," than they might have been a century or two ago. For one thing, commodification is not necessarily as closely associated with appropriation, in the usual senses of the word, as it used to be. To take one instance, the commodification of price changes (or, more precisely, of price change certainty and uncertainty) involved in today's vast market in futures does not involve the appropriation of price certainty in any conventional sense. Nor is appropriation sensu stricto involved in the widespread practice of short selling, or "shorting." Rather, securities are only borrowed, to be sold when the price is high and bought back when the price is low, even though not only the securities but also the underlying assets are also thus woven into expanded networks of exchange and thus, arguably, intensified commodification. As a whole, complex financial derivatives are types of commodities that are only tenuously related to seizure or assertion of property or access rights. The special powers they exert over land, water, and air are tangible and attributable in a sense to expanded commodification, but they are not achieved through standard processes of appropriation. The commodification of pollution presents another example. To avoid "takings" lawsuits from business that could result from governments' tightening emissions restrictions under cap-and-trade systems, tradable pollution permits are generally claimed in legislation *not* to be property rights of any kind — in spite of being universally treated as assets and commodities. Hence, when European corporations are granted monetizable rights to dump greenhouse gases in the atmosphere or to use foreign vegetation or soil to soak up their carbon dioxide emissions, something is being appropriated, but that appropriation is hedged about and governed in novel ways.

Bakker's criterion of standardization, similarly, falls short of capturing some of the most significant innovations in post-1970 commodification. Standardization is a process best applied to things that, to borrow the useful phrase of Donald MacKenzie (2009), have already been, at a basic level, "made the same." For example, it was only because wheat was already a universally recognized classification that the Chicago Board of Trade in the nineteenth century was able to formulate practices for isolating categories such as "No. 2 spring wheat" as standardized commodities in an era in which grain was being transformed from a product that stayed in sacks from farm gate to final buyer into a "golden stream" coursing through railroad cars and grain elevators (Cronon 1991, 97–147). No such preexisting classifications exist in the burgeoning post-1970 trend of ecosystems services commodification. The immediate challenge of commodification here is not in standardization but in making things the same in the first place. So-called carbon markets, for instance, despite having been in existence for two decades, have yet to identify an intelligible or universally agreed-upon thing to trade in; the tradable unit is typically defined, as Jillian Button (2008, 581) observes, "not in terms of *what* the unit is, but what it entitles the holder to *do*." Tradable carbon permits allow their buyers to emit greenhouse gases, but whether the permits are to be defined as access rights to global carbon-cycling capacity, whether this or that type of counterfactual reductions (emissions below "what would have happened otherwise") of different greenhouse gases can be accepted as exchangeable equivalents, whether units from countries with different emissions caps should be treated as the same, and so forth are matters of unceasing controversy, as will be described below. Indeed, as the markets expand, carbon commodities "created to fit the necessities of a market system" tend to become "increasingly vague" rather than globally standardized (Rosales 2006, 1046; see also Munden Group 2011). As Jessica Dempsey (2011, 199–203) points out, prerequisites for standardization are even harder to entrench in biodiversity markets, where the relative "clarity" of carbon commodities, ironically, is viewed with envy. Similarly, although wetlands bankers have been trading wetlands certificates since the 1980s, they not only have "not settled upon a system of measurement" but also have "not even agreed upon what the commodity is that they wish to measure" (Robertson 2004, 367). Standardization is also unattained (and probably unattainable) in contingent valuation, cost-benefit analysis, and the other types of "proxy commodification" (Castree 2003) that have enjoyed such a resurgence during the neoliberal era and that attempt to set up replicable and verifiable practices of market-like valuation ab ovo in circumstances in which none of the customary webs of market practices, with their constraining and enabling features, yet exist (Lohmann 2009). This is to say nothing of the growing range of commodification processes in which the very notions of commensuration and standardization are problematic, such as in markets for art or the more bespoke range of speculative financial products (Karpik 2010; Cooper 2010).

In an important survey article, Castree (2003) wisely sidesteps many of the difficulties of pat definitions of "commodification" by instead offering a somewhat "thicker" account featuring various conditions seen by scholars of a Marxist bent as normally required for it. Castree's work suggests the fruitfulness of approaching commodification not as a phenomenon to be corralled by a sharply rounded-off dictionary entry but as a subject of an open-ended, discursive, dialectical effort to grasp the nature of contemporary crises. Castree (2003, 278) notes, for example, that the mere quality of being exchangeable has long been, for Marxists, "too thin a basis to specify what is entailed by capitalist commodification" as a process in which "qualitatively distinct things are rendered equivalent and salable through the medium of money," with particular use values commensurated and acquiring the "general quality of exchange value." Yet even this specification is not nearly enough, for most Marxist scholars, to get a real handle on the subject. Further elements identified by Castree include privatization, which is as much about "control over commodities—prior to, during and after exchange—as it about ownership in the technical, legalistic sense" (279; cf. Bakker 2007a); alienability or "detachability" from sellers; individuation against a background of legal and material supporting contexts; functional abstraction, or the separating off of a measurable characteristic of a thing or process from the thing in its original context; spatial abstraction, or the treating of an individual thing in one place as the same as something (ostensibly different) elsewhere, as when wetlands in place A are mobilized to "replace" wetlands in place B (Robertson 2000); commensuration with other commodities in a way that allows a thing or process to function as one moment in the accumulation of capital; and displacement or fetishization, according to which commodities appear as things rather than socionatural relations. Elsewhere, Castree and others have expanded this treatment still further by linking new types of commodification of nature to neoliberalism or financialization (Smith 2008; O'Connor 1994a; Castree 2008a; Heynen et al. 2007; Moore 2010). Again, the point is not to arrive once and for all at a "master definition"—neoliberalization and financialization are themselves contested shorthands for complex processes about which controversy is rife—but to suggest practical tools for investigators whose instinct is that the study of commodification is especially important at the current moment. Initiatives such as Castree's thus offer a useful way station or orienting device between, on the one hand, misleading abstractions and, on the other, lengthy "thick descriptions" (Geertz 1973) of particular instances of commodification whose relevance to other cases or to broader historical trends usually requires some effort to tease out.[1]

This chapter is intended as a limited further contribution to the effort to come to terms with neoliberalism's new "nature commodities" by mediating between, as it were, dictionary entries and encyclopedias, abstract definitions and thick descriptions. While concerning itself with a single aspect of the commodification of a single "ecosystem service"—climate stability—it does so by proposing, and demonstrating the use of, a conceptual instrument conceivably applicable to a variety of problematic post-1970 commodities. Focusing largely on the moments of "making things the same" that are crucial for so many such commodities, this instrument consists of summarizing complicated practices of commensuration in thematic *performative equations*, around each of which specific accounts of actors, methodologies, institutions, resistances, and outcomes can then be collected. *Equations* are used simply because they are a tidy way of expressing the relations of "sameness" that most commodities require for their operation. These equations are *performative* (Austin 1961; MacKenzie 2006) in the commodification context in the sense that, rather than being true or false descriptions of entrenched states of affairs, they constitute commitments to help bring about the equivalences they specify.

The Object of Climate Commodification

The question of *why* attempts have been made for the last two decades to make commodities out of climate, and how this was made possible by earlier twentieth-century developments, is beyond the scope of this chapter, although it has been discussed elsewhere (Lohmann 2006, 2011). The question here is *how* one might fashion these commodities. The answer is not immediately obvious. Global warming results mainly from the transfer of carbon from a fossil pool locked underground to a separate pool circulating above the ground among the atmosphere, oceans, vegetation, soils, freshwater, and surface rocks. This transfer is irreversible over humanly relevant time scales. It follows that sustaining—or "producing"—the use value of a liveable climate requires keeping remaining fossil fuels in the ground.

To put it another way, given path dependence (Arthur 1999) and the way that fossil fuels have become "locked in" (Unruh 2000, 817) to industrialized societies' ways of life, climate reality calls for political mobilization behind immediate long-term investment programs in new, nonfossil energy, transport, agricultural, and consumption regimes, particularly in the North. Climate reality also requires programs for shifting state subsidies from fossil fuels to existing initiatives defending or constructing low carbon means of livelihood. Above all, that reality demands widespread alliance building in support of the social movements that are already directly or indirectly addressing the below- to above-ground transfer of carbon. These include movements working to "keep oil in the soil, coal in the hole, and tar sand in the land" in the Niger Delta, Alberta, Ecuador, South Africa, Appalachia, and elsewhere; stopping the development of dozens of coal-fired power plants in the United States, Britain, India, Thailand, and other countries; fighting agrofuel projects whose effect would be to sustain a transportation infrastructure designed for oil; and working to ban banks from supporting fossil-intensive or fossil-extractive projects. Increasingly, such movements are aligning themselves with those in support of ecological and peasant agriculture, more democratic public health and energy provision, cleaner air and water, and an end to militarism, environmental racism, and extractivism.

Prima facie, a climate change mitigation commodity would need to support movement building of this radical kind. Yet how might it be possible to buy and sell contributions toward the long-term political shift away from fossil fuels that such movements are working toward? In a tongue-in-cheek but nonetheless instructive proposal, legal scholar Douglas Kysar (2010) suggests that the "legal and political actions" that have "dramatic impact" on historical trends would have to be commodified. The resulting products could be sold by, for example, "indigenous groups that entirely block new exploration activities" or "forest dwelling communities that successfully fight to stop logging." Investment banks seeking to craft new financial products would "devote themselves . . . to the identification and promotion of critical political interventions by disempowered voices for sustainability." Accumulation would be a matter of investing in instruments that maximized structural societal change over the long term.

To make accounting, ownership, and capital accumulation possible, Kysar's climate commodity would have to turn the qualitative relations that make up movement building and historical process into quantitative ones. But obstacles would arise immediately. For example, consumers would need to know, and producers to guarantee, what increment of historical change toward a halt to fossil fuel extraction each commodity sale represented. But who would quantify the extent to which each unit of the commodity contributed to undoing the social complexities of fossil fuel path dependence, and how? If different units contributed different increments of historical change depending on the particular pathway they were aggregated into, and the paths were incompatible, how would the units be commensurated, much less standardized? How would the historical effects of private ownership on the dialogue and movement building comprising the "labor" producing the climate commodity be calculated? If the expert storytellers (Beckert 2011) whose services would be needed to help price the commodity attached a particular value to rolling back the dominance of a rampant financial sector, would Goldman Sachs sell the associated securities? The only way of removing such difficulties for accumulation would be to demote the market to being a provider of unspecified and unquantifiable "climate services"—in which case it would lose most of its usefulness for policymakers and its appeal to other potential customers.

An Alternative Model

The alternative to the immediate, dizzying multiplication of paradoxes of Kysar's whimsical proposal is to construct a commodity based on the enclosure and commodification of pollution sinks, whose extent the state defines in terms of limits on the quantity of molecules that can be emitted. This is what the United States' sulphur dioxide trading system, instituted in the 1990s, did, and it is the model followed by the Kyoto Protocol's carbon market, the EU Emissions Trading Scheme, and all other actually existing climate markets.

The advantages are clear. Molecules can be counted (in many pollution markets, a metric ton is the unit of measurement). Molecules come "prestandardized" in the sense that they are the same the world over. Molecules—or molecular motions—can also be laid claim to, and so, at least in principle, can the sinks that absorb molecules—for example, oceans, trees, or lands that soak up carbon dioxide. Quantifiability and ownability make it possible to buy and sell rights to emit CO_2—essentially, rights or access to the earth's carbon-cycling capacity in the oceans, atmosphere, soil, vegetation, and rock. And with quantifiability, measurement, and property claims comes, too, the possibility of systematized market exchange and large-scale accumulation. Focusing commodity construction on molecular rather than social movements, in addition, has the advantage of tapping into an existing cultural and political momentum. Even before ecosystems markets became all the rage, the issue of "global climate change" had become identified with the largely molecular "concerns that have guided climate modelers in their daily practices" (Goeminne 2012, 3). Modelers' efforts to build reliable climate knowledge from enormous amounts of disparate data had in turn been enabled partly by a more general postwar institutional movement centered on prediction and forecasting that has also profoundly shaped formal economics (Mirowski 2011). A molecular approach to climate change both reinforces and is reinforced by widespread contemporary "processes of depoliticization" as well as fetishistic and apocalyptic disavowals of the "multiple and complex relations through which environmental changes unfold" (Swyngedouw 2010, 214, 220).

As is the case with the mechanics of any commodification process, however, the flip side of the advantages of the choice of object is a complex set of costs and resistances (Marx [1867] 1976, 198). Both these advantages and the "overflows" (Callon 1998a) that are their inevitable counterpart can be mapped and analyzed according to the open-ended set of constructed equivalences or performative equations that, along with various technologies, persons, institutions, disciplines, and bits of nonhuman nature, make up one part of the infrastructures of markets (Callon 1998a; MacKenzie 2006). The particular set of equivalences that symbolize and form part of the infrastructure for climate commodity formation are sketched out in the remainder of this chapter, forming an analytical backbone around which the logics and resistances associated with climate commodification can be taxonomized and discussed.

Instead of founding themselves on the premise that action on climate entails keeping fossil fuels in the ground, then, climate services markets are based on the equation:

$$\text{a better climate} = \text{a reduction in } CO_2 \text{ emissions}$$

An immediate effect of this institutionally and politically convenient choice of object is to entrench a process that continually reframes the climate problem in ways that disentangle it from climate history and the transfer of fossil fuels out of the ground and reembed it in neoclassical economics, chemistry, and a variety of other quantitative disciplines. Eliding the multiple differences between reducing emissions and addressing the climate crisis, the foundational equation above obscures, for example, the difference between stepwise molecule reductions over the short or medium term and actions that integrate into a program that would result in most remaining fossil fuels being left in the ground permanently.[2] In addition, it ignores the nonlinearity and unverifiability of the relationship—a consequence of the physically "chaotic," flip-flop nature of the atmospheric system—between any given increment of reduction, on the one hand, and any given increment of climate benefit, on the other. Also elided is the difference between molecules that can be classified as "survival" emissions and those that can be classified as "luxury" emissions (Agrawal and Narain 1991)—an elision that has climatic as well as class consequences, since "survival" emissions tend to have causes, dynamics, and historical accompaniments different from those of "luxury" emissions. In such effects lie the seeds of a whole spectrum of resistances to climate commodification, ranging from the criticisms of many climate scientists and environmentalists to the opposition of grassroots social movements in the global South.

Molecular Equations and Their Discontents

Once molecular flow management is made into the object of political action on climate, then the fact that CO_2 molecules are identical throughout the world implies that the following equation can also be made into a guiding principle of climate policy:

stopping transfer A of x molecules of CO_2 into the atmosphere = stopping transfer B of x molecules of CO_2 into the atmosphere

So, too, then, can its corollaries:

stopping the transfer of x molecules of CO_2
into the atmosphere in place A =
stopping the transfer of x molecules of CO_2
into the atmosphere in place B

stopping the transfer of x molecules of CO_2
into the atmosphere through technology A =
stopping the transfer of x molecules of CO_2
into the atmosphere through technology B

stopping the transfer into the atmosphere of
x molecules of CO_2 of underground fossil origin =
stopping the transfer into the atmosphere of
x molecules of CO_2 of surface biotic origin

Such equations mark practices that allow firms, investors, and speculators to benefit from cost differentials between various investments in reduced molecule flows. If it is cheaper to invest in mandated reductions in place A than in place B or in reductions that use technology A rather than technology B, then the choice will be obvious for any business; it will be similarly obvious if it is cheaper to invest in forest conservation than in technologies that use less fossil fuel. Hence the celebrated cost-saving "flexibility" of climate markets in which one "reduction" can be traded for another in what proponents hope will be a maximally liquid trading system.

The molecular focus of the four equations displayed above gives them the rhetorical or mythical (Zbaracki 2004) power of chemistry. Who could deny that molecules of CO_2 are the same whatever their origins and locations? There is, however, once again a flip side: the appearance of indisputability is achieved only by reframing a question of climate history as one of chemistry. In reality, it can make a difference to the trajectory of global warming whether a given reduction in CO_2 flows is attained in place A or place B, through technology A or technology B, or through industrial restructuring or forest conservation. Equating CO_2 reductions that result from different technologies makes it not only possible but often necessary to make climatically wrong choices in the name of molecule prices—for example, to reduce molecule flows through routine, cheap efficiency improvements that entrench coal use and delay long-term nonfossil investment, or to build destructive hydroelectric dams that do nothing to displace coal and oil, rather than to select no-carbon technologies that form an integral part of a long-term program for phasing out fossil fuels (Driesen 2003; Taylor 2012). Equating reductions in place A with place B, meanwhile, obscures a number of geographically specific factors that make a difference to energy transitions, including the greater influence on technology development that a reduction in emissions from a particular industrial process might have in a high income country, where it is more expensive, than in a low income country (Alfredsson 2009). It can also make a difference whether an identical reduction is achieved through technological innovation or halting forest degradation. The traditional objection both inside and outside United Nations climate negotiations to policies that rely on trees for "reductions" is that they weaken incentives for structural change in industrialized societies. This quality is especially important given two further realities: first, that no increase in biotic carbon on the earth's land surface would be capable of keeping out of the atmosphere and the oceans more than a fraction of the comparatively enormous stores of fossil carbon now being transferred to the surface from underground; and second, that the delays in the inevitable decarbonization of industrial societies enabled by exchanging biotic for fossil carbon make that decarbonization rapidly more expensive, and thus more daunting, over time. In short, as equations of chemistry, the four equations displayed above are true; as equations of climatology, they are false; but as equations that help structure market exchange, they are perhaps best regarded as neither true nor false but rather as normative expressions of, and commitments to, novel commensuration practices that are unavoidably conflict-ridden and uncompletable. Their truth-value in terms of chemistry is relevant to their performativity in the market context only insofar as it provides them with some moral cachet in a context in which the climate issue has already been molecularized. Their truth-value (i.e., their falsity) in terms of climatology is relevant to their performativity only insofar as it tends to undermine their credibility among those who insist that climate markets should be about climate.

The four equations above give rise to other types of "blowback" as well. For example, making cost per molecule the criterion of choice between technology A and technology B helps pave the way for land-intensive (and thus socially discriminatory) programs that attempt, at least ostensibly, to "replace" fossil fuels. Among these are strife-ridden agrofuel schemes in countries such as Brazil, Honduras, and Indonesia, as well as wind power projects such as those in Mexico's Tehuantepec isthmus, where indigenous communities have regretted cheaply signing over land to private wind farm developers from Spain and Mexico who profit not only from electricity sales but also from trading the resulting pollution rights in Europe or using them to sustain their own fossil-fueled installations. By abstracting from the tendency for pollution to be concentrated in what in the United States are called "poorer communities of color," technology and place neutrality also help ground future capital accumulation in historical patterns of class and racial discrimination, ensuring staunch opposition to carbon markets from networks of underprivileged communities ranging from the California Environmental Justice Movement (California Environmental Justice Movement 2010) to India's National Forum of Forest Peoples and Forest Workers.[3] Equating fossil and biotic carbon intensifies climate class struggle in the same way, since doing so provides additional economic and "scientific" sanction for extensive landgrabs from the poor (Gregersen et al. 2010; Leach, Fairhead, and Fraser 2012), whose livelihoods are likely to come into competition with carbon-absorbing projects and who may also see their store of knowledge of low-carbon subsistence livelihoods depleted as a result (another effect that is inconvenient to include in carbon calculations). The "cost curves" that the equation makes possible also tend to abstract from the difference between forest clearing for commercial agriculture, on the one hand, and rotational forest farming that involves subsequent regrowth of forests and storage of carbon, on the other. This abstraction both works against long-term forest conservation and, again, facilitates the deskilling of forest dwellers. As Nathaniel Dyer and Simon Counsell (2010, 69) comment, the "argument that we need a new economic model to account for [climate change] externalities and to put our economies on a sustainable path" has ironically led to cost curves that, with their "hidden costs and partial analysis," are "similar to the narrow economic approach that contributed to the problem that we are now attempting to solve." Thus Aritana Yawalapiti, an indigenous leader in the upper Xingu River region of Brazil, reported in November 2010 that carbon forestry promoters visiting his territory had told his community that they would have to reduce forest burning if they were to be paid for producing carbon pollution licenses. But, Aritana objected, "We always burn at a place where we fish, hunt or open a small farmland area. . . . [W]e open a space to farm, we plant, we collect manioc, after some years everything recuperates again. . . . [T]he forest grows back, while we plant at another place."[4]

In sum, the cost advantages of "geographical neutrality," "technology neutrality," or "carbon source neutrality" map onto various aspects of "mission drift" in climate markets as an instrument of environmental policy, as well as a number of other severe market-undermining effects. Of course, contradictory effects following from the abstraction involved in commodification are nothing new: in the nineteenth-century Chicago grain markets, for example, commodity abstraction, while making futures possible, also engendered the possibility of, for example, market-crippling speculative corners or conflict over profits that elevator operators gained simply because they were located in a position that enabled them to mix grain from many different farmers in order to minimize the quality of each bulk consignment they sold within a standard grade (Cronon 1991, 134ff.). Questions regarding climate services markets' contradictions, however, tend to be of far greater scope than those that challenged the new Chicago wheat market of the 1850s onward. For example, to what extent have the abstraction processes involved in the formation of climate commodities undermined their ostensible policy purpose altogether? Can the commodities even be made coherent enough to survive? To what extent can their self-undermining dynamics be brought out of the "black box" in which they are currently concealed in United Nations and neoliberal environmental discourse and in the work of academics and other experts? The more carefully the performative equations structuring climate commodities are unpacked, the more salient such questions become.

Offset Equations and the Attack on Nonexpert Agency

One further step in this unpacking process involves examining the equations that structure the practices responsible for creating what are known as "offsets." Under the Kyoto Protocol carbon market, as well as the European Union Emissions Trading Scheme and other climate market arrangements, polluters subject to government emissions caps, as well as funds, banks, or other private or public enterprises, can finance carbon-saving projects outside the caps and use the resulting extra pollution rights—offsets—in lieu of emissions reduction obligations, or sell them to third parties, or speculate with them. Thus:

$$CO_2 \text{ reduction under a cap} = \text{offset outside the cap}$$

For example, European Union Allowances (EUAs), the emissions permits traded under the EU cap, are exchangeable with Certified Emissions Reductions (CERs), which are Kyoto Protocol carbon offsets generated in southern countries outside the European cap:

$$EUA = CER$$

Offsets thus make possible additional abstractions from place and widen the scope of possible molecular cost savings from technology choice or forestry. That is, they take the "spatial fix" (Harvey 2006a) of cap and trade (which moves pollution around a "capped" landscape to wherever it is cheapest to abate) one step further, to territories not covered by caps, especially the global South, where carbon cleanup is cheaper (Bond 2010). This multiple boundary-crossing function is reflected in the distinctive equation in which offsets are embedded:

$$\text{reduction under a cap} = \text{"avoided" emission outside the cap}$$

This equivalence allows offset projects that emit greenhouse gases (and most do) to license the emissions of still more greenhouse gases elsewhere—as long as they emit less than "would have been released" in the absence of carbon finance. For instance, carbon traders or capped polluters in the UK can purchase carbon pollution rights from highly polluting sponge iron factories in India, provided the factory owners can convince UN regulators that technological improvements have resulted in less CO_2 than would have been emitted otherwise and that this saving is measurable according to approved criteria.[5] The cost savings are considerable. In September 2012 the price differential between cheap CERs and more expensive EUAs on the Bluenext spot market in Paris was US\$7.52—a gap that can also be profitably exploited by speculators.

In order to arrive at a single amount of "carbon saved" in India that can be priced and substituted for measured and verified industrial emissions reductions in the UK, however, a single counterfactual story line must be posited as a baseline. Methodologically, then, the offset equation requires that counterfactual history be given the same epistemic status as actual history:

$$\text{actual } CO_2 \text{ reduction} = \text{counterfactual } CO_2 \text{ reduction}$$

"What would have happened" in the absence of carbon credit sales must be treated as determinate and quantifiable in the same way that CO_2 reductions under a cap are determinate and quantifiable. These equations commit offset creators and traders to a deterministic modeling of human and nonhuman actors, since only on deterministic assumptions is it possible to isolate the single story line required for commodity pricing, based on the starting conditions of a counterfactual without-project scenario. This commitment to recasting political debate about alternative futures as disputes about the correctness of technical predictions has affinities with a more general postwar technocratic dedication to ideals of forecasting and apolitical control, with rational actor theory in economics, and with a more recent trend in the financial markets toward "mechanized" storytelling about the future through mathematical models, whether those models are used as confidence-building devices (Beckert 2011), technologies of a new, credit ratings–dominated pattern of investment (Ouroussoff 2010), or actual engines of mass production of certainty commodities (Tett 2009). Not surprisingly, it is subject to similar blowbacks. One is simply that the methodological ambition is too high. As George Soros (2008) and many others have emphasized for the financial markets, calculative technologies, when pushed beyond a certain point, undermine their own efficacy. Kevin Anderson (2012) of the Tyndall Centre for Climate Change research makes a similar point about carbon calculation: "The offsetters' claim to account for carbon leakage over the relevant timeframe presumes powers of prediction that could have foreseen the internet and low-cost airlines following from Marconi's 1901 telegraph and the Wright brothers' 1903 maiden flight. Difficult though it is for contemporary society to accept, ascribing any meaningful level of certainty to such long-term multiplier effects is not possible and consequently offsetting is ill-fated from the start." This indeterminacy underlies part of the prolonged methodological agony of conscientious offset accountancy experts such as Michael Gillenwater (2012), who asks, "What does it mean for an offset project to be real? What would an unreal offset project be? How could we tell if it was unreal, and is this something we should be concerned about?"

A second contradiction is that, necessary as a deterministic model is for offset calculation, it is also necessary that the technical experts and investors responsible for offset projects be exempted from it: the offset commodity form requires that they be rewarded for making a choice in what otherwise is an unalterable course of history. Offsets, that is, must attribute agency to privileged actors while denying it to every other human or nonhuman agent. This is, of course, a move familiar from the annals of colonial and postcolonial history, as well as of neoclassical economic theory. But the denial of workers' and farmers' capability to create their own history is no more likely to escape resistance in the present case, where it is closely integrated with commodity formation in climate markets, than it is elsewhere. Early on, for example, one group of Brazilian activists denounced the "sinister strategy" of claiming that a pig iron industry was creating emissions reduction "equivalents" by burning plantation charcoal rather than coal:

> What about the emissions that still happen in the pig iron industry, burning charcoal? What we really need are investments in clean energies that at the same time contribute to the cultural, social and economic well-being of local populations. . . . We can never accept the argument that one activity is less worse [*sic*] than another one to justify the serious negative impacts. . . . [W]e want to prevent these impacts and construct a society with an economic policy that includes every man and woman, preserving and recovering our environment. (FASE 2003)

CO_2 Equivalence and the Pitfalls of "Efficiency"

Among its many other advantages, climate markets' focus on molecules opens up the cost-saving possibility of using greenhouse gases other than carbon dioxide in the formation of climate commodities. Here market construction has benefited from the work of the Intergovernmental Panel on Climate Change (IPCC), which, prompted by the UN's need for national greenhouse gas accounts as well as its own molecular preoccupations, has attempted to commensurate CO_2 with a range of other greenhouse gases such as methane (CH_4), nitrous oxide (N_2O), and various chlorofluorocarbons and fluorocarbons, including the industrial by-product HFC-23 (IPCC 1996), according to their relative effects on global warming, or global warming potential (GWP). The result is the following equations:

$$CH_4 = 21 \times CO_2$$

$$N_2O = 310 \times CO_2$$

$$HFC\text{-}23 = 11,700 \times CO_2$$

These equations can then be used to elaborate a climate commodity in terms of CO_2 equivalent (CO_2e) rather than just CO_2. Having abstracted from the climate crisis to CO_2 molecules, in other words, climate services markets now abstract from CO_2 and other gases to posit portmanteau quasi-molecules of CO_2e, which assume and expand the fetish status already accorded to CO_2. In the performative equations previously analyzed in this chapter, accordingly, "CO_2" can often be replaced with "CO_2e" (depending on the particular market's rules), amplifying each equation's scope.

The consequence is to make the trade in climate services enormously more "efficient" and profitable, both for fossil fuel users and for dealers in pollution permits, due to the cost savings achieved by substituting new molecular "raw materials" for carbon dioxide. For instance, given that burning off just one ton of CH_4 can generate salable rights to release twenty-one tons of CO_2 in Europe, it is not surprising that—to take one example—more than two dozen giant hog farms operated by Granjas Carroll de México, a subsidiary of the US-based Smithfield Farms, have sought extra revenue by capturing the methane given off by the huge volumes of pig excrement they produce, burning it, and then selling the resulting carbon credits to Cargill International and EcoSecurities. Merely by destroying a few thousand tons of HFC-23, similarly, the Mexican chemical manufacturer Quimobásicos is set to sell over thirty million tons of carbon dioxide pollution rights to Goldman Sachs, EcoSecurities, and the Japanese electricity generator J-Power (UNEP, Risoe Centre 2010). Assuming that destruction of HFC-23 can be carried out for US$0.25 per ton of CO_2e and that a ton of UN offset pollution rights can command US$3.11 on the European Union Emissions Trading Scheme (EU ETS) spot market (at historically low September 2012 prices), both the company and the financial sector intermediaries it sells to can realize good profits. Industrial buyers of the permits can in turn save over US$140 per ton by using the rights in lieu of paying fines for not meeting their legal emissions requirements. Today, the cleanup of HFC-23 and N_2O generates more profit for their manufacturers than the primary products of the processes in question (Pearce 2010), creating perverse incentives to make global warming worse (Szabo 2010; Schneider 2011). Such industrial gas offsets—generated at a handful of industrial installations in China, India, Korea, Mexico, and a few other countries—still account for the bulk of Kyoto Protocol carbon credits. The CO_2-equivalent construct also makes possible many other creative climate commodity-producing schemes. Coal mines in China, for example, can now produce and sell carbon credits by burning off some of the methane that seeps out of underground veins on the grounds that by converting methane into carbon dioxide, the projects do less damage to the atmosphere than would have been the case otherwise.

An additional advantage of the GWP construct is that it facilitates the running together, in a seemingly self-evident way, activities with different effects on climate history. Thus ex–World Bank executive Robert Goodland (Goodland and Anhang 2010, 11), noting that "domesticated animals cause 32 billion tons of carbon dioxide equivalent, more than the combined impact of industry and energy," can effortlessly draw the politically convenient conclusion that "replacing livestock products with better alternatives would . . . have far more rapid effects on greenhouse gas emissions . . . than actions to replace fossil fuels with renewable energy."

One of the contradictions of the pursuit of efficiency through CO_2 equivalents, as with that of the fossil-biotic carbon equivalance, is, accordingly, an inbuilt bias against the rural poor that has already generated criticism from activist networks such as La Via Campesina and the World Rainforest Movement. But the problems and resistances go a great deal deeper. For example, devising the performative equations about different greenhouse gases displayed above requires a great deal of fudging, leading to continuing technical disputes. Each greenhouse gas behaves qualitatively differently in the atmosphere and over different time spans, and the control of each has a different effect on fossil fuel use. The IPCC itself winds up revising its calculations of the CO_2-calibrated GWP of various gases every few years and insists on giving gases different GWPs over twenty-year, one-hundred-year, and five-hundred-year time horizons. But even such token caveats cannot be accommodated by a market that requires a single, stable number in order to make exchange possible. The UN carbon market, for example, disregards the IPCC's recent revisions in GWP figures, discards twenty-year and five-hundred-year figures, and ignores the often enormous "error bands" specified by the IPCC (in the case of HFC-23, plus or minus five thousand CO_2 equivalents). Again, translation and simplification turn out to have heavy blowbacks.

Ownership and "Deresponsibilization"

If there is to be a market in CO_2 emissions reductions, someone must "produce" them, and someone must buy them. To put it another way, if there is to be a market in greenhouse gas pollution dumps, someone must make them scarce, someone must "own" them, and someone must "rent" them. Setting up this apparatus can only be the job of governments, which must impose both the need for reductions (by making pollution dumps scarce) and the means of "producing" or owning them. Governments achieve the former by imposing "caps" or limits on emissions on companies or economic sectors. To accomplish the latter (i.e., create a reduction commodity), governments need an equation:

regulated reduction of CO_2 emissions to level c within time period p = tradable right to emit CO_2 up to level c by the end of period p

Carbon dioxide reductions (and, by inference, climate action) can accordingly be achieved by the production of tradable pollution rights whose scarcity or otherwise is determined by government fiat. Progressive carbon dioxide reductions can in turn be achieved by relying on an additional equation:

reducing CO_2 emissions progressively through regulation = issuing fewer tradable rights to emit CO_2 in period $p + 1$ than were issued in period p

The producers or owners of these rights are, in the first instance, governments themselves. European Union Allowances, for example, are "produced" in preset amounts by the pens or keystrokes of politicians and bureaucrats under the European Union Emissions Trading Scheme. They are then sold or, more usually, given away free to large private-sector polluters. Assigned Amount Units (AAUs), one of the climate commodities of the Kyoto Protocol carbon market, are meanwhile "produced" by conferences of the parties to the UN Framework Convention on Climate Change before being distributed, again free of charge, to the national governments of industrialized countries.

In helping to "perform" climate commodities, the above two equations at the same time engender additional severe and contradictory overflow effects. First, equating reductions with salable property rights once again distances the new markets from their assigned function in climate policy. As fossil fuel use becomes more deeply entrenched through a "polluter earns" system, the preoccupation with price discovery draws emphasis away from the long-term structural change demanded by global warming. All things being equal, corporations will choose cheaper alternatives, but if long-term structural alternatives have not been made available, not even the highest prices can compel anyone to choose them; on the contrary, they are likely to incite revolts against the trading system's design. Nor have low prices ever historically been drivers of the kind of structural change that global warming demands. The EU ETS has not incentivized investment away from fossil fuels even in the one sector, electricity generation, that has been consistently short of emissions rights (see, e.g., Deutsche Bank 2009).

Second, the performative equations above embed, in the institutions surrounding climate markets, a far-reaching capillary system of practices that, at all levels, deresponsibilizes industrial societies with regard to global warming. For example, instead of being fined for exceeding Kyoto Protocol emissions targets (which, as Herbert Docena [2011, 42] points out, implies the commission of an offense), industrialized country signatories are encouraged to buy extra pollution permits from abroad to compensate for their failure (an action that connotes the acquisition of an entitlement). At the same time, in Nigeria, the Philippines, South Africa, Guyana, and many other southern countries, governments are incentivized by carbon markets not to promulgate or enforce environmental laws (which attribute responsibility for harm to defendants) but instead to allow their societies to remain dirty in order to be able to sell pollution rights from subsequent cleanup programs. Increasing institutionalization of opportunity-cost estimates in the design of biotic offset schemes, similarly, favors the relatively wealthy—those with the means to destroy forests wholesale—over poorer communities that follow a more environmentally benign approach, thereby further reducing the space for practices that work to recognize and gauge responsibility for destruction or preservation (McAfee 2012a; Lang 2011). Tens of thousands of experts, traders, bankers, lawyers, accountants, consultants, and bureaucrats working in a US$100 billion–plus global market setting fuel emission proxy factors, commenting on carbon project design documents, formulating schedules and criteria for payments for forest conservation certificates, making submissions to UN carbon market regulators, hedging investments, buying land, tallying molecules, balancing accounts, establishing ownership, and discovering prices continually produce and reproduce deresponsibilization in each of the offices and arenas they work in. Rich nations are thereby "transformed" from climate offenders or debtors into climate leaders or benefactors. Colonialist ideologies temporarily challenged by the early 1990s global debate over climate change have been rehegemonized not so much through propaganda, moral reasoning, bad science, or outright threats and bribes as through the repetition and accretion of thousands of quotidian technical practices surrounding commodity construction and operation. Accompanied as it is by the erosion of juridical approaches to the environment and the reduction of fines to fees, this colonialist resurgence has, unsurprisingly, provoked strong opposition to the new climate commodities from social movements and activists in both northern and southern nations (Osuoka 2009; Docena 2010).

Conclusion: Regulation and Internalization

The strenuous commodifying processes of simplification, abstraction, quantification, propertization, and so forth reflected in performative equations constitute the deep structure of the attempted "internalization of environmental and social externalities" that is one face of the market environmentalism characteristic of the neoliberal era. These processes continually reinterpret and transform the challenges they confront; their goals are never exogenous but are incessantly reshaped by the very process of addressing them. This chapter has argued that, with respect to the climate crisis in particular, internalizing externalities through commodity formation, however profitable the result, constantly gives rise to fresh externalities that are so overwhelming that, from an environmental point of view, they invalidate the project.

From this perspective, the commonly heard appeal to "regulation" as a solution for such failures needs disambiguation. Does regulation mean revising, elaborating, and extending the contradictory performative equations that provide infrastructure for the new commodities in question, as is implied by most critical writings on climate markets (e.g., Newell and Paterson 2010; Perdan and Azapagic 2011; Bumpus 2011)? Or does it, rather, mean progressively "deactivating" some or all of the equations? The burden of this chapter has been that, in the case of climate services markets, progressive deactivation will be the environmentally wiser approach in view of the incessantly ramifying counterproductivities that any variants of the relevant performative equations are bound to engender.

For example, no additional equivalences, surveillance procedures, or technical criteria for determining when a carbon offset project goes beyond "business as usual" could ever be capable of relieving the contradictions built into the equation:

$$\text{actual } CO_2e \text{ reduction} = \text{counterfactual } CO_2e \text{ reduction}$$

On the contrary: given the equation's commitment to the impossibilities of verifiable counterfactual history, they merely give these contradictions "more room to move," to quote a resonant phrase of Marx ([1867] 1976, 198). The effect has been to reinforce the supply-side dominance in the offset markets of large polluting corporations that operate in the global South—Sasol, Mondi, Rhodia, Tata, Birla, Jindal, and the like (UNEP, Risoe Centre 2010)—which are better able than others to devote resources to navigating the growing regulatory and planning mazes that the contradictions feed in the service of gleaning new revenues for activities that reinforce fossil fuel use. That, in turn, signifies another step backward in the struggle over climate change.

One virtue of breaking down the omnibus category of commodification into bite-sized chunks using the tool of performative equations is that to do so gives concrete content to the observation that commodification and decommodification have many forms and degrees, as well as a spectrum of different types of internal structures. To do so also provides a criterion for distinguishing instances of regulation—of whatever motivation or provenance—that contribute toward a goal of decommodification from those that do not. Such a criterion can be useful for climate activists in deciding which tactics to adopt, since even governments that have subordinated their climate policies to a commodity framework are sometimes induced to undertake actions with modest decommodification effects that, if supported, may lead to larger and more constructive changes. For example, the EU decided in 2011 to stop applying the equation

$$\text{HFC-23} = 11{,}700 \times CO_2$$

by banning HFC-23 credits from sale as of 2013. The reasons for this move were complex, involving not only scandals over the issuance of a flood of blatantly bogus pollution rights from industrial gas projects (EIA 2010) but also fears that European industries in the sector in question may relocate to the global South to take advantage of offset revenues, a desire to reduce transaction costs in the manufacture of carbon offsets by sourcing them from entire sectors rather than individual projects, and worries that an oversupply of carbon credits will undermine market operations. Nevertheless, the curb does demonstrate the possibility of rolling back commodification rather than extending it, as do environmentalist campaigns to abolish offsets and hence deactivate equations such as:

$$\text{EUA} = \text{CER}$$

Breaking down specific neoliberal nature-commodification processes using open-ended sets of performative equations, then, is one way of teasing out a core of both analytic and practical strategic sense in reactive slogans such as "our Earth is not for sale" as well as in overly abstract academic definitions of commodification. In clarifying contemporary struggles over market environmentalism, it may help identify and expand spaces for potential alliances among various movements questioning commodification—whether of climate, water, electricity, health services, ideas, biodiversity, or genes—and supporting land and labor rights, alternative energy and transport, food sovereignty, and public control of the financial sector.

Notes

This chapter has benefited from discussions with and comments from Oscar Reyes, Steve Suppan, Andres Barreda, Jutta Kill, Ricardo Coelho, Hendro Sangkoyo, Martin Bitter, Esperanza Martinez, Ivonne Yanez, Matthew Paterson, Silvia Ribeiro, Raul Garcia, John Saxe Fernandez, Herbert Docena, Patrick Bond, John O'Neill, Erik Swyngedouw, Mark Schapiro, Wolfram Dressler, Rob Fletcher, Bram Büscher, and friends at the Centre for Research on Socio-Cultural Change at the University of Manchester. Much of the material has been previously published in *Capitalism Nature Socialism* and the *Socialist Register*.

1. One of the best models of such "thick descriptions" remains Cronon's treatment of wheat and lumber in the nineteenth-century US Midwest, but there are of course dozens of other valuable studies, including Thompson (1990), Thomas (1991), Mirowski (2011), Sivaramakrishnan (1999), and so on.

2. The difference between the two is illustrated by the fact that the industrial slowdown resulting from the financial crisis of 2007–8 resulted in more CO_2 emission reductions than all the world's climate markets put together had achieved (Chaffin 2010), yet has not changed structural dependence on fossil fuels.

3. See *Mausam: Indian Climate Change Magazine*, 2008 and 2009, NESPON, Kolkata, http://www.thecornerhouse.org.uk/sites/thecornerhouse.org.uk/files/Mausam_July-Sept2008.pdf and http://www.thecornerhouse.org.uk/sites/thecornerhouse.org.uk/files/Mausam2–5.pdf.

4. R. Sommer, 2010, video interview with Pirakuma Yawalapiti, Xingu spokesperson, about carbon trading, Xingu River, Brazil, http://www.youtube.com/watch?v=_JSM6gaM9CA and http://www.youtube.com/watch?v=JMs3szvzfeA&feature=related.

5. Similarly, forest carbon projects can generate carbon credits even if they allow an increase in deforestation, as long as the increase is "less than would have happened otherwise" (see, e.g., American Carbon Registry 2011).

PART III

Nature on the Move

The Global Circulation of Natural Capital

CHAPTER EIGHT

Nature on the Move I

The Value and Circulation of Liquid Nature
and the Emergence of Fictitious Conservation

Bram Büscher

This chapter is part of a broader project to understand the place of conservation in the critical analysis of the relations between nature and contemporary capitalism. While there are vast literatures on how "nature" and "capitalism" interrelate, these are overwhelmingly geared toward the manner in which the latter *uses*, *transforms*, and/or *impacts upon* sociobiophysical natures. A solid theoretical framework for thinking about the place of the *conservation* of nature within contemporary capitalism is still embryonic. This is odd, considering that the fate of modern conservation has been interwoven with capitalist trajectories since its inception in the eighteenth and nineteenth centuries (Grove 1995). In fact, the preservation of the world's "last wild places" appears as a classic Polanyian double-movement, a direct response to the alienation of humans from nature and the massive transformation of nature under capitalist expansion (Cronon 1996). At the same time, by separating rural people from their land, conservation aided in the formation of the labor force that industrial capitalism needed (Perelman 2007) while proving a valuable tool in colonial administrative control (MacKenzie 1997). More recently, an intensive and pervasive proliferation of protected areas has accompanied the rise of neoliberal capitalism since the late 1970s (Brockington, Duffy, and Igoe 2008), while the 1990s and 2000s have given rise to popular paradigms such as "payment for ecosystem services" and novel approaches such as biodiversity derivatives, wetland credits, species banking, and more (Robertson 2004; Cooper 2010; Sullivan 2012b). All these are based on the assumption that capitalism and conservation are—can be made—compatible (see Brockington and Duffy 2010b), which leads to a pertinent question: How can we understand the conservation of nature as a capitalist project?

This question is the topic of a nascent but swiftly growing literature. Igoe, Neves, and Brockington (2010), for example, focus on how a Gramscian hegemonic "historic bloc" intersects with an economy focused on Debordian "spectacle" to produce the *idea* that capitalism and conservation can indeed be compatible (see also Fletcher 2010 for a poststructuralist perspective). While these authors convincingly show how in this way the prediction by green Marxists that the "second contradiction of capitalism" would lead people to demand ecosocialism (O'Connor 1998) has been neutralized—or delayed—they leave implicit the question how the conservation of nature actually functions *as capital* in the twenty-first-century global economy. Over the last two decades, this question has become a prominent one, particularly after the recent (or ongoing) financial crisis. Not only has the idea that business should "green" itself received a massive boost, the financial crisis also led to calls for a "global Green New Deal" and a "green economy" that focus on shifting the global political economy from extractive to nonextractive or nontransformative use and its concomitant valuation of nature and natural resources (Büscher and Arsel 2012).[1] We thus witness the capitalist system increasingly accepting the effects of the "second contradiction" yet trying to deal with it by making it part and parcel of the system, by giving "value" to the conservation of nature. It does this in the only way it knows how to give things value: by taking them up as commodities in capital circulation, by finding new ways to guarantee "nature on the move."

Obviously, this makes sense from the perspective of capital. After all, capital, according to Marx ([1867] 1976, 256), is "money in process," "value in process." If anything, the last years have again made abundantly clear that when capital stops moving, the system in which it thrives is in deep crisis. Hence, all over the world, governments were fixated on getting money moving again and so turn it back into capital. Similarly, in our times of multiple environmental crises, we see many actors working hard to turn the conservation of nature into capital so that it can take its "rightful" place in global markets and no longer be dispensed with as mere "externality." This leads to a further dilemma: how does "conserved nature"—what I will call "liquid nature"—circulate as capital, as "value in process," and what does this mean for the value of nature?[2] This is a significant question with potentially quite radical implications for (neo- or post-) Marxist theory and for conservation.

Let me briefly outline why before moving on to discuss the question in more depth. Most fundamentally, the commodities "produced" by capitalist conservation (aim to) turn "production" on its head and hence ingrain ideas about (the production of) value. The accepted, Marxist way of thinking about the relation between capitalist production and nature goes something like this: "Human beings exploit nature in all sorts of ways. It hardly seems possible to imagine otherwise. The transformation of nature, though it takes place under all manner of conditions and through all manner of socially embedded practices, is an absolute requirement for the production of anything" (Henderson 2003, 77). Of course, this is generally correct, with one major possible exception, namely, *when capital seeks to produce the nontransformation of nature*, most especially through its conservation. Now, it has to immediately be added that the conservation of nature does not mean the nontransformation of nature. The opposite is true: nature is actively produced and transformed through its conservation (Brockington and Duffy 2010b; Dressler 2011). Yet, the manner of production and transformation is rather different from what is generally understood as the "transformation of nature under capitalism." It is a transformation that aims to leave nature (materially) unexploited and unused and is as such seen as diametrically opposed to and—importantly—*fit to offset* "traditional" production processes that do (materially) exploit and use nature. Phrased differently, the *value* in this product, at least theoretically, is found exactly in the fact that nature is (believed to be) not (materially) used, transformed, or exploited.

In contemporary conservation, this idea has become known under the banner of "natural capital," which provides "environmental services" to humans. Nature-to-be-conserved functions in this rhetoric as a peculiar kind of *fixed capital* whose value circulates through the capital embodied in and implied by its environmental services. I refer to this as liquid nature—nature made fit to circulate in capitalist commodity markets—the potential for which, I argue, has been made possible within a change in the nature of circulation in contemporary capitalism. Yet, these services, like the land and nature they are derived from, are a form of fictitious capital: "capital without any material basis in commodities or productive activity" (Harvey 2006a, 95). In Marxist terms, this would also mean that they cannot hold any value, as they have not been (directly) produced through human labor. Given this, the question "how does conserved nature circulate as *capital*, as value in process" has potentially fundamental implications for ingrained ways of thinking about value, nature, and the relations between production and circulation in capitalism. Indeed, a central argument of this chapter is that the analysis of conserved nature as capital necessitates a shift in emphasis from production to circulation. *It is (the nature of) contemporary capitalist circulation that enables the circulation of liquid nature as a form of fictitious capital, the ultimate result and consequence of which is "fictitious conservation."* This, however, is not to discount production. To the contrary: production, as we will see, remains crucial, but quite different from "standard" Marxist theories of production.

In what is to follow, this argument is approached from two angles. First, I will outline the nature of circulation in capitalism and how this has changed over the last three to four decades. Next, I will discuss how this transformation relates to attempts to enable the circulation of nature, leading to the argument that to make markets for conserved nature fully liquid — or to create fully liquid nature — capital has had to "elevate" nature from fixed to fictitious capital. The difference is that in the latter case, the link between actual natures and their conservation through digitalized financial mechanisms is severed, so creating "fictitious conservation." The penultimate section discusses the notion of fictitious conservation in more depth and explores its consequences for Marxist theories on production, circulation, and value. The chapter ends with some brief concluding thoughts.

Before moving on, it is important to emphasize that all of this is not a matter of mere abstract political economy: to make liquid nature believable, legitimate, and manageable, capital has had to and continues to create particular governmentalities and associated ideological belief systems. These matters, however, are outside the purview of this chapter and will be taken up by Jim Igoe in his companion piece. Moreover, it also does not mean that no alternative ontologies and epistemologies exist when it comes to "nature on the move" and that these could potentially provide ways out of the current capitalist deadlock. These will be discussed by Sian Sullivan in her companion piece. The sole objective of this chapter is a stepwise *theoretical* exploration of how conserved or liquid nature becomes capital that circulates with great speed in our contemporary global economy. It is an exercise in logical reasoning, not an empirical investigation, although the potential empirical and practical implications might be considerable.

The Nature of Circulation in Capitalism and "Fictitious Capital"

The ensuing discussion on the nature of circulation in contemporary capitalism will start by going into some "fundamentals" of capitalist circulation based on Marx's *Capital* ([1867] 1976) and Harvey's *The Limits to Capital*

(2006a). I will then move beyond this "deep structure" of capitalism to incorporate how circulation has changed alongside recent changes in global capitalism. Hence, I explicitly *start* with Marx, not *end* with his work, as is sometimes the case in Marxist-inspired work. I will argue that several aspects of Marx's work will need to be reconsidered and/or expanded in order to fully understand *contemporary* capitalist circulation that has made liquid nature possible.

The basis of capitalist circulation for Marx ([1867] 1976, 227–28) starts when commodities are "sold not in order to buy commodities, but in order to replace their commodity-form [C] by the money-form [M]" and when "the change of form becomes an end in itself." This leads to the famous conversion from C–M–C to M–C–M, whereby a capitalist "throws money into circulation, in order to withdraw it again by the sale of the same commodity" (249). Money thus becomes "money in process" or "value in process" and therefore capital. This has due implications: "The circulation of money as capital is an end in itself, for the valorization of value takes place only within this constantly renewed movement" (253). When capitalist circulation becomes an end in itself, and under the pressures of competition, the "immanent laws of capitalist production" start confronting "the individual capitalist as a coercive force external to him" (381). A tremendous amount of faith is thus placed in the (seemingly) "exogenous" process of circulation to keep accumulation on track. As even mainstream economists recognize, however, this is obviously incorrect. In the endless complexities of the differentiated circulation and realization times of capital, production, commodities, and values, it is clear that circulation in the aggregate is never an even, consistent, or automatic process (Marx [1870] 1978). If circulation of capital converged exclusively around commodities, capitalism would quickly become immensely unstable. This imminent instability is, for Harvey (2006a, 254), why credit is vitally important to the system.

While full discussion of credit is beyond the scope of this chapter, some remarks are important for clarifying its focus on circulation. Harvey (2006a, 285) talks about the "immense potential power that resides within the credit system": "Credit can be used to accelerate production and consumption simultaneously. Flows of fixed and circulating capital can also be co-ordinated over time via seemingly simple adjustments within the credit system." Credit, however, leads to what Marx called "fictitious capital," which Harvey (2006a, 95) describes as "money that is thrown into circulation as capital without any material basis in commodities or productive activity." In turn, he argues that "the potentiality for 'fictitious capital' lies within the money form itself and is particularly associated with the emergence of credit money" (267). He explains as follows:

> Consider . . . a producer who received credit against the collateral of an unsold commodity. The money equivalent of the commodity is acquired before an actual sale. This money can then be used to purchase fresh means of production and labour power. The lender, however, holds a piece of paper, the value of which is backed by an unsold commodity. This piece of paper may be characterized as *fictitious value*. Commercial credit of any sort creates these fictitious values. If the pieces of paper (primarily bills of exchange) begin to circulate as *credit money*, then it is fictitious value that is circulating. A gap is thereby opened up between credit moneys . . . and "real" moneys tied directly to a money commodity. . . . If this credit money is loaned out as capital, then it becomes *fictitious capital*.

While arguing that credit can function to stabilize circulation, Harvey (2006a, 288) adds that this does not mean that credit solves capitalism's inherent contradictions. Indeed, it embodies the contradictions it aims to solve, but on new levels and with new complexities:

> What started out by appearing as a sane device for expressing the collective interests of the capitalist class, as a means for overcoming the "immanent fetters and barriers to production" and so raising the "material foundations" of capitalism to new levels of perfection, "becomes the main level for over-production and over-speculation." The "insane forms" of fictitious capital come to the fore and allow the "height of distortion" to take place within the credit system. What began by appearing as a neat solution to capitalism's contradictions becomes, instead, the locus of a problem to be overcome.

Once a process of relying on debt to guarantee and intensify accumulation has been set in motion, there is no way back: accumulation has to continuously increase in order for "fictitious capital" to retain its "value." The use of credit thus adds a major impetus to ensure that capital is truly "money in process" or "value in process" and thus that the velocity of circulation must continuously increase. Circulation, Marx ([1858] 1973, 255) remarked in the *Grundrisse*, "has to be mediated not only in each of its moments, but as a whole of mediation, as a total process itself." What this points toward is that a certain velocity of circulation helps sustain a particular amount of "fictitious capital" and how with its further institutionalization capitalism becomes progressively dependent on the circulation and proliferation of this type of capital.

Much has changed since Marx's day and even since Harvey first published *The Limits to Capital* in 1982. It is thus necessary to account for subsequent dramatic changes in the global political economy and their effects on capitalist circulation. This is crucial, since while Marx's and Harvey's analyses point us in the right direction, one thing both these scholars did not foresee is the way in which global capitalism would (try to) adjust in relation to the environmental degradation it engenders. This was obviously not a major issue in Marx's time, but even Harvey does not devote much attention to this in his work and so completely misses the important connections between changes in contemporary capitalism and the energy expended to finding ways to green capitalism through conservation (Büscher et al. 2012).

The background to these changes is found in a central imperative of capitalism, namely, "to reduce the time and cost of circulation so that capital can be returned more quickly to the sphere of production and accumulation can proceed more rapidly" (Smith 2008, 126). On a global scale, Castells (2000, 136–37) argues, this has truly become possible with the advent of new information and communication technologies, including "advanced computer systems" that allow "new, powerful mathematical models to manage complex financial products, and to perform transactions at high speed." In this process, "the whole ordering of meaningful events loses its internal, chronological rhythm, and becomes arranged in time sequences depending upon the social context of their utilization" (492).

So far, so good, but an apparently irreducible obstacle to this dream of unfettered hypercirculation remains. For as Smith (2008, 126) further argues, "the circulation of value requires also a physical circulation of material objects in which value is embodied or represented" (see also Henderson 2003, 43). Understanding how capitalism may be transcending (or perhaps circumventing) this apparently irreducible obstacle requires further theorization of value and circulation. Let us start with LiPuma and Lee (2004, 19), who make the same point about the central imperative of capitalism as Smith but draw more radical implications about circulation: "The basic or founding argument is that the internal dynamic of capitalism compels it to perpetually and compulsively drive toward higher and more globally encompassing levels of production. This directional dynamic has engendered such progressively ascending levels of complexity that connectivity itself has become the significant sociostructuring value, leading to the emergence of circulation as a relatively autonomous realm, now endowed with its own social institutions, interpretative culture, and socially mediating forms." While the level of "autonomy" can be debated, the fact is that connectivity has become a "significant sociostructuring value," to the extent that Boltanski and Chiapello (2007) have elevated this value to the center of their analysis of the "new spirit of capitalism." LiPuma and Lee (2004, 97), however, draw their conclusions about circulation from their analysis of financial derivatives, which for most of their history "were production-focused and functionally geared to hedging." This changed in the 1970s with several far-reaching "institutional changes and the liberalization of national capital controls" (98). As a result, "the essential movement of the market was away from hedging on production to wagering on circulation" (99). Next, LiPuma and Lee describe how this process started leading a life of its own to the extent that it has created a system "in which means dominate ends": "The goal of financial circulation increasingly shapes the means of its realization" (154).

Again, some elements of LiPuma and Lee's (2004, 179–80) overall analysis can be debated, most especially the power they attribute to financial capital in the West and their "move away from production," as it is clear that financial capital has recently reemphasized material production, particularly land and agricultural commodities in the global South, resulting in massive landgrabs (Borras et al. 2011). That said, it is undeniable that the direction of change in global capitalism has been toward unleashing financial markets and hence massively increasing the intensity and velocity of capital circulation (Moore 2010; Marazzi 2011). What, then, does this mean for the concept of value?

LiPuma and Lee (2004, 83), again, take a radical step, arguing that "standard macroeconomic theories of international trade and exchange rates, or Marxist approaches that originate from a labor theory of value, appear to have little to say about circulation." Technically, this is not correct: many do have many things to say about circulation, but they interpret this rather differently. The central question here, at least from a Marxist perspective, is where and how value is produced. In this chapter, I follow Phil Graham (2007, 174), who argues that while Marx's theory of value still forms the "deep structure" of capital, contemporary notions of value (e.g., those embodied by financial derivatives) are no longer the ones that Marx first articulated:

> Today it is not the muscle-power of people that provides the most highly valued labor forms. Far more intimate aspects of human activity have become technologized and exposed to the logic of commodification.

Correspondingly abstract forms of value have developed. Value production, in turn, has become more obviously "situated" in the valorized dialects of "sacred" and powerful institutions, such as legislatures, universities, and transnational corporations. In official political economy, value has moved from an objective category that pertains to such substances as precious metals and land, to become located today predominantly in "expert" ways of meaning and, more importantly, in their institutional contexts of production.

This has major consequences for the nature of circulation in capitalism. It means that capital increasingly circulates as "expert ways of meaning" and "institutional contexts of production," for example, through reports, policy briefs, think tanks, brands, marketing, and so on (Goldman and Papson 2006), but also through financial derivatives, futures, and other financial constructs (Lee and LiPuma 2002). In other words, what circulate mostly these days are forms of *fictitious capital*—capital that does not directly have "any material basis in commodities or productive activity" (Harvey 2006a, 95). In addition to credit, this capital takes the form of a whole host of financial and nonfinancial derivative "products" that, among others, focus on institutional or organizational efficiency; management of meaning; technological, informational, and communicative "innovation"; or simply speculation. These "products" all crucially depend on a concept of value that is ephemeral and transient. Indeed, Graham (2007, 4) argues that "what we call 'values' are more or less ephemeral products of evaluation," which, "like all aspects of meaning, . . . are socially produced and mediated."

This, it must be emphasized, is not to say that production-based labor is not important. It does mean that its role in the production of value has changed, most notably through a shift in emphasis toward circulation, in that circulation increasingly determines production rather than the other way around (see also Marazzi 2011, 48–49). LiPuma and Lee (2005, 424, emphasis added) articulate these changes as follows: "We appear to be . . . heading into an era where speculative capital, a socio-historically specific concept of risk and derivatives products have become the centre of the financial clockwork that turns the hands of contemporary capitalism. *There is thus reason to believe that circulation-based risk represents a new self-structuring dynamic that is superimposed upon and structurally supersedes an earlier form grounded in production-based labour.*" Circulation superseding and determining production, however, is not new, as pointed out for California agriculture in the late nineteenth and early twentieth centuries by Henderson (2003). What has changed over the last decades, or so I argue, is that *the valorization of production is increasingly alienated from the act of production.*

This, of course, has consequences for production in general and for the "production of conservation" under capitalism specifically. Production in general, in this process, is relegated to producing "underlying assets" for the (financialized) derivative structure that is the prime focus for value creation in contemporary capitalism.[3] In turn, it is in this context that we see global capitalism increasingly directing its attention to dealing with its negative environmental consequences in a way that mediates its worst excesses while opening up new frontiers for capital accumulation (Arsel and Büscher 2012). To enable this process, several fundamental changes in the way capitalism operates and generates value are necessary, first, and most especially, to value the nonuse or nonextraction of nature (and hence paying for labor that conserves rather than appropriates or destroys nature) while simultaneously trying to reduce the "physical circulation of material objects" that Smith (2008, 126) argues is necessary for the circulation of value and, second, to replace these with creating the possibilities for the circulation of liquid nature as capital. It is to these changes and their challenges and critiques that we now turn.

The Circulation of Liquid Nature as Capital

Anno 2014, it is abundantly obvious that our planet's natural environments are being transformed and commodified with unprecedented intensity and speed. As policymakers, NGOs, businesses, and politicians work to alleviate the growing concerns about capitalism's negative ecological record, they often do so under the banner of "natural capital" (see Costanza et al. 1997). This (usually) involves bringing nature deeper into contemporary capitalism through mainstream neoclassical economic tactics (Burkett 2005, 113). Nature as "capital," in this discourse, appears to function according to classical forms of *fixed capital*, which "circulate as value while remaining materially locked within the confines of the production process" (Harvey 2006a, 209). This is achieved in large part through the products nature creates, namely, a whole host of different "environmental services" (Sullivan 2009).

What different variations of the idea of environmental services have in common is their rather simplistic presentation of how embedded value is "transported" from the producing entity "nature" to the consuming entity "humanity." These variations, according to proponents, could be different categories of services, including supporting, provisioning, regulating, and cultural ecosystem services (Millennium Ecosystem Assessment 2005, vi).[4] The exact nature of these different types of services, however, is not relevant; what matters for the analysis is that a complex array of services is tied to a range of "constituents of well-being" (vi) through a valuation model that relies on *monetary* payments in order to assign quantitatively comparable values to qualitatively incommensurable conditions and relationships (Kosoy and Corbera 2010). Arguably the most important policy result of this thinking is the currently trendy payments for environmental services (PES) paradigm.

Of course, the standardization of value measures is an extremely complicated process, requiring a great deal of speculation by those doing the "measuring" and "valuing." In this section I will not focus on precisely how this is done. Rather, based on the two functions of money, namely, "as a measure of value and as a medium of circulation" (Harvey 2006a, 292–93), my primary concern is, first, to briefly outline the implications and problematic aspects of the monetization of nature and, second, to discuss how nevertheless this monetized nature is supposed to become *circulating* and *valuable* global conservation capital.

Importantly, if nature is expressed in money, we need to first clarify our conceptualizing of "nature," particularly if some kind of *material, biophysical* nature is to be conserved through some kind of commodified, *abstract* value circulation. Biodiversity conservation is explicitly *not* interested in what Castree (2003, 286) calls "internal nature," nature that has been brought almost entirely under human technological control, like genetically modified seeds. It *is* explicitly interested in nature that "still retains the independent capacity to act," or what Castree calls "external nature." Although most external nature is "inherently social" (Smith 2007, 77), fundamentally shaped by human thought and action, it remains far more unruly and encompassing than internal nature. It is precisely this kind of unruly and encompassing nature that biodiversity conservation sets into motion so that it may circulate as a form of fictitious capital.

To theorize this circulating nature, it is necessary to account for both the biophysical and social aspects of nature and to engage with them as interconnected and mutually constituting realms (Castree 2000; Carolan 2005; Büscher et al. 2012). After all, as argued by Smith (2007, 33), "capital is no longer content simply to plunder an available nature but rather increasingly moves to produce an inherently social nature as the basis of new sectors of production and accumulation." However, as Carolan (2005, 400, 409) cautions from a critical realist position, it is also necessary to maintain some distinction between these categories such that they do not wind up simply merging into one another. He thus distinguishes three categories: Nature, nature, and "nature."[5] The first is "the Nature of physicality, causality, and permanence-with flux," the second is nature as sociobiophysical phenomenon, and the third is "nature" as discursive construction. While all three are important, in this chapter I am centrally concerned with the latter two categories, their intersections and mutual constitutions, in the circulation of conserved nature as capital. Conservation is always to a large extent a struggle between different "natures" in terms of "discourse, power/ knowledge, cultural violence, and discursive subjugation" (401). As these discursive regimes influence human action, they play an active hand in shaping biophysical nature (Carrier and West 2009). At the same time, biophysical nature shapes, limits, and defines discursive regimes of "nature," such that the two are in constant dialogue, as shown by Igoe's and Sullivan's companion pieces.

This brief discussion has obvious implications for the circulation of conserved nature as fictitious capital. If it is to circulate in the capitalist economy, conserved nature must be monetized. If it is monetized, it will be expressed and understood in quantitative terms, which erases the "ontological depth" and qualitative complexity of relationships between Nature, nature, and "nature." Specifically, as Burkett (2005, 122–24) elaborates, it is possible to identify five important problems with the monetization of nature: (1) "unlike money, 'nature cannot be disaggregated into discrete and homogenous value units'"; (2) a reliance on money leads to "inadequate accounting for the irreversible character of many natural processes" (e.g., there is no reason to assume that the monetary value of an ecosystem will go up before its depletion/extinction is irreversible); (3) monetization involves an absolute "tension between money's quantitative limitlessness and the limits to natural wealth of any given material qualities"; (4) "the price of a resource stock is not determined solely by its absolute size" but by many other aspects of how markets work, meaning that "price may not rise as depletion occurs"; and (5) "higher resource prices may actually accelerate a resource's depletion by spurring technological advances that reduce extraction costs and/or lower the amount of the resource needed per unit of final goods, thereby encouraging its further use to increase total output." Burkett (115, emphasis added) concludes that even "many ecological economists have resisted it [*natural capital*] on the grounds that it is irreparably anti-ecological" and "lends a spurious legitimacy to the commercialisation of nature and *its reduction to a productive input*."

These points highlight the problematic and contradictory effects of transforming nature into a quantitative, monetary input—a point I will come back to below. At the same time, these criticisms have not prevented many conservation, business, and government actors from trying to monetize nature. In fact, it has spurred them on even more (Bracking 2012; MacDonald and Corson 2012). In this endeavor, they have been enabled, I argue, by the contextual transformations in global capitalism laid out in the previous section, most notably, the proliferation of complex forms of fictitious capital, changes in the production of value, and how these have influenced interrelated processes of production, consumption, and circulation. In other words, while the idea of monetizing ecosystem services as the product of "fixed" natural capital is a problematic and, critics would argue, futile and false solution, *it is only the starting point* for those who aim to bring conserved nature into contemporary capital circulation. They need to go further still and find ways to link capitalist conservation to a political economy where value has become ephemeral and located "in 'expert' ways of meaning and, more importantly, in their institutional contexts of production" (Graham 2007, 174).

And this is exactly what has been happening, as shown by recent scholarship on conservation and capitalism. Thus, Garland (2008, 67) has posited a "conservationist mode of production" that "lays claims to natural (and thus fixed) capital" and adds value to it "through various mediations and ultimately transform it to a capital of a more convertible and globally ramifying kind." Brockington (2008) chronicles the "power of ungrounded environmentalisms" by emphasizing how conservation celebrities enable (mostly Western) audiences to reestablish their bonds with the wild through commodified representations of nature. Igoe (2010) records how conservation produces and turns upon Debordian "spectacle" in the "global economy of appearances," particularly how spectacular media representations of nature are dominating the way environmental nongovernmental organizations communicate and "sell" their conservation messages. Dressler (2011), based on research in Palawan Island, the Philippines, notes how capitalist conservation shifted from first to third nature, a nature that lives up to how tourists would like nature on Palawan to be. Lastly, I have earlier shown how conservation initiatives around the 2010 soccer World Cup in South Africa produced and incorporated what I call "derivative nature," the systemic preference on the side of capital for idealized representations of nature and "poor locals" in order to attract tourists and investment (Büscher 2010b). What these disparate examples have in common is that they show how contemporary conservation fundamentally adheres to and relies on "ephemeral values" to enable the circulation of conserved nature in contemporary capitalism.

Having stated this, it is crucial that we do not take this argument too far: just as a rapidly circulating and speculative financial realm ultimately still depends on a more "mundane" production, distribution, and consumption of asset streams (Leyshon and Thrift 2007, 98), so is contemporary conservation still deeply intertwined with the material realities of sociobiophysical nature. This, for instance, is clear from work by Katja Neves (2010, 721), who shows that the commodity fetishization of whale watching is not as diametrically opposed to exploitative whale hunting as it imagines itself to be. In fact, she argues that the "transition from one to the other is more closely related to transformations in the global capitalist economy than to enlightened progress in human-cetacean relations." The new production of conserving whales through ecotourism, then, precariously links making audiences literally buy into commodified and romanticized whale encounters and shielding them from the negative material sides of the same, for example, the disturbance of whale ecology and carbon-packed air travel. This poses a more general problem, namely, that the circulation of conserved nature as capital has to be achieved through creating "derivative" ephemeral value while at the same time remaining inextricably linked to material (sociobiophysical) nature.

In other words, for conserved nature to truly function as capital, it has to go beyond environmental services. After all, the generally accepted definition of PES talks about a "well-defined environmental service" that is sold by a particular provider to a buyer "if and only if the ES provider secures environmental service provision (conditionality)" (Wunder 2005, 3). The "problem" here is that this does not necessarily involve competitive markets and indeed often comes down to mere "compensation schemes." True capitalist marketization of conserved nature would need to go far beyond this in order to link material nature with ephemeral values. In business terms, most environmental services markets lack sufficient "liquidity." Liquidity is business lingo for a market with an ever-ready supply of sellers and buyers where assets can easily be bought or sold with little effect on price levels. It means that commodities need to be fully "alienable" and/or fully transferable at minimum transaction cost. This presents fundamental problems for markets of "environmental services," as their liquidity is usually circumscribed in space and time (see also Fletcher and Breitling 2012). Thus when the rather naive idea of PES has scarcely become popular in mainstream conservation, it is already being overshadowed by a host of much farther-reaching proposals to turn conserved nature into circulating capital. We are currently witnessing the creativity at work of those who push the frontiers of capitalist commodification ever further, as conservation derivatives, "sustainability enhanced investments," wetland and mitigation banking, biodiversity offsets, and other schemes are rapidly making headway in conservation and extraconservation arenas.

While an extensive discussion of these separate schemes is neither possible nor necessary here (see Sullivan 2012b), what they have in common is that risks related to, impacts on, and incentives toward biodiversity (conservation) are financialized and subjected to market exchange. Mandel, Donlan, and Armstrong (2010, 45–46), for example, promote "conservation derivatives" as hybrids of "two types of financial instruments" "in which an insurance derivative is issued with modifications to allow responsible action to decrease the likelihood of the insured event." Wetland and mitigation banking and biodiversity offset schemes, in contrast, are geared toward offsetting the impact of development projects by (at least) restoring or reviving the same amount of biodiversity that was destroyed by the project (see, e.g., http://bbop.forest-trends.org/ and Robertson 2000 for a critique). Taken together, the goal of all these mechanisms is to make markets for conserved nature more fully liquid, which indeed is how it is referred to in practice.[6] Let us now look at the implications of this development on Marxist theory and conservation in more detail.

The Emergence of Fictitious Conservation

The *ultimate* objective of getting market liquidity right is of course the lubrication of producing greater surplus value or profits.[7] The *immediate* objective of liquidity is to facilitate faster and/or smoother turnover of capital and thus to increase the velocity and/or stability of capital circulation. The Platonic ideal of liquid nature is one in which monetized forms would be completely free from the material contexts and relationships that produced them. In reality, of course, "financial superstructures" are always entangled in material realities (Leyshon and Thrift 2007, 98). Neoliberal conservation's entanglements with material realities are the topic of another emerging body of literature and need not detain us (but see West 2006; Neves 2010; Büscher 2010b). What is important to note here is that these entanglements occur in "a world that can no longer be directly grasped" (Debord 1967, 11), in which production and consumption have become so separated that "their relationship becomes all but unfathomable, save in fantasy" (Comaroff and Comaroff 2002, 784).

Accordingly, the connections and disconnections between consumers of liquid nature and the conditions and relationships that produced it have become so complicated that they are, for most intents and purposes, severed. It is not just that individual producers, consumers, and natures are no longer directly in touch, though this is often certainly the case. The point is that the various products derived from many distinct natures have to become standardized and utterly abstracted in order to be exchangeable. This is not just a *strategic* process, as Smith (2007, 29) has it; it is a *necessary* one. This is achieved in large part through *securitization*: the standardization and rationalization of "nontransparent and localized commodities . . . so that different buyers and sellers in different places around the globe can understand their features and qualities and exchange them easily" (Gotham 2009, 357).

Hildyard (2008, 4–5) takes the idea of securitization one step further, arguing that it is "a process whereby assets that generate regular streams of income . . . are sold to a newly created company (a Special Purpose Vehicle [SPV] . . .). The SPV then issues derivatives . . . that give investors the right to the income stream from the assets." As these highly complicated processes are stacked on top of one another, one can immediately see how they completely erase any local, qualitative, and spiritual properties and contexts around an "environmental service" through their subjection to utterly abstract numbers on marketized value indices. This has resulted in the profoundly "new face of nature," depicted in figure 8.1.

Proponents of the marketization of conserved nature usually argue that securitization helps stabilize and balance markets and prices. Yet, examples from other markets that depend on the "liquidization" of fixed capital commodities reveal this is not the case. Taking the housing market, which had such a major role in the financial crisis, as an example, Gotham (2009, 357, 368) contends that "the housing finance sector is permeated by significant contradictions and irrationalities that reflect the disruptive and unstable financial process of transforming illiquid commodities into liquid resources" and that this "conceptualization of securitization as a process of creating liquidity out of spatial fixity dovetails with theoretizations that emphasize the conflictual, contested and deeply contradictory nature of uneven geographical development." This is a stark warning for ecosystem markets. Fundamentally, it points to the ways in which securitization artifices have systematically transformed homes and neighborhoods into fictitious capital that can circulate in the global economy without concern for, or even knowledge of, the material and social conditions that produced them. I am arguing, by extension, that similar securitization artifices are systematically and fundamentally separating liquid forms of conserved nature from the material and social conditions that produced them. The upshot is the full-fledged conversion of conserved nature into capital, thereby enabling its

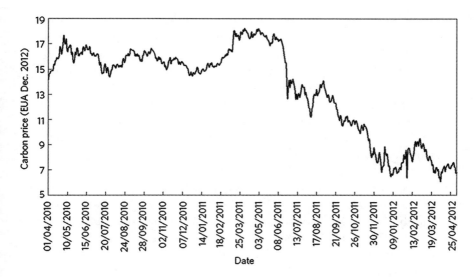

Figure 8.1. The "new face of nature," or a typical index for pricing (here carbon). Source: http://www.ecosystemmarketplace.com, accessed September 22, 2010.

ultimate purpose: becoming a new vehicle for money in process, or value in process. Conservation, in other words, has become fictitious capital, which leads to what I call "fictitious conservation," conservation without any direct basis in material, sociobiophysical nature.

Fictitious conservation has not displaced or subsumed more traditional forms. Rather, it accompanies them, intertwines with them, and infuses them with its logic in ways remarkably analogous to interactions between "nature" and nature as outlined above. Traditional forms of conservation may continue to protect animals, landscapes, and ecosystemic processes. Increasingly, however, *the valorization of these activities is alienated from them and subject to broader processes of the circulation of liquid nature.* At the same time, the logic of fictitious conservation is increasingly geared toward the production of liquid nature tout court. In losing much of its basis in sociobiophysical nature in favor of liquidity, the idea of "fictitious conservation" can almost be taken literally; after all, how can conservation alleviate the "second contradiction" of capitalist expansion if it is capitalist expansion that is the ultimate objective to begin with?

The implications of all this are legion. For one, it adds an additional layer of complexity to Smith's (2007, 33) cogent discussions of "nature as accumulation strategy," in which he argues that the "horizontal integration of nature into capital" (the exploitation of material nature) is now being complemented by the "vertical integration of nature into capital" through the "production of nature 'all the way down'" and "its simultaneous financialisation 'all the way up.'" A focus on the circulation of liquid nature further complicates this picture. Liquid nature, I have argued, depends on a conceptualization of ephemeral value that blurs Smith's horizontal and vertical axes of nature as capital beyond recognition. It moves through these intermittently and simultaneously as a frenzied circulation of a seemingly integrated "nature" and nature.

The analysis also complicates Smith's (1996; 2007, 25) discussions of the "production of nature" as well as Garland's (2008) concept of the "conservationist mode of production." While I agree with Smith's epistemology behind the idea of the production of nature as taking both material and discourse seriously, I believe that conserved nature as capital in the context of contemporary capitalism emphasizes that "formerly distinct spheres of analysis"—production, distribution, consumption, and circulation—are converging more than this thesis can give credit for (Graham 2007, 7). Being overly "productivist" can blind analyses for "other processes that simultaneously socialize nature" (Castree 2000, 285) while it also obscures the ephemeral and hybrid character of value in contemporary hypercapitalism.[8] Likewise, the concept of a "conservation mode of production" cannot do justice to the ways in which nature and conservation are increasingly becoming "valuable" in the global economy, namely, as fictitious capital, which depends on the ever-increasing velocity of circulation.[9] Nature is not only produced. It is constantly on the move, along with fictitious versions of the very forces that produced it, through simultaneous and intertwined processes of circulation, consumption, distribution, and production.

Yet, while having said this, the analysis at the same time leads us to the argument that the emphasis in the creation of value has shifted from production to circulation. The Marxian theory of value would stress that value is ultimately produced through the surplus extracted from labor in production, which in turn happens through the appropriation of nature. This becomes problematic, of course, when environmental services circulate as fictitious capital without having been produced by human labor. In fact, the idea of capitalist conservation says that humans should be paid to *forgo* the creative appropriation of nature. As such, capitalist conservation is at the same time an acknowledgment of production and its role in the transformation of nature, as well as its (hoped-for) negation. These two opposites, in turn, are brought together in the idea that natural capital commodities (seem to) skip the phase of material production to focus on the *production of circulation*. Central in all of this is the elimination of the (traditional) role of labor and hence the questioning of what Hannah Arendt ([1958] 1998, 85) referred to as "the glorification of labor as the source of all values." In other words, the point of capitalist conservation becomes giving (ephemeral) value to the elimination of labor's appropriation or transformation of nature.

Interestingly, Arendt in the 1950s had already criticized Marx in a similar way. In *The Human Condition* she argues that Marx's conceptualization of labor as being directly embedded in the life process through the metabolism of nature leads to a "fundamental and flagrant contradiction" in his value theory (1998, 103–4). She argues, on the one hand, that "when Marx insists that the labor 'process comes to its end in the product,' he forgets his own definition of this process as the 'metabolism between man and nature' into which the product is immediately 'incorporated,' consumed, and annihilated by the body's life process." On the other hand, she insists that "while it was an 'external necessity imposed by nature' and the most human and productive of man's activities, the revolution, according to Marx, has not the task of emancipating the laboring classes but of emancipating man from labor; only when labor is abolished can the 'realm of freedom' supplant the 'realm of necessity.'" Interestingly, the capitalist system is now trying something similar: to emancipate capital circulation from labor and its role in the transformation of nature as a way of "offsetting" other labor processes that do (need to) continue to transform nature. This, of course, is inherently contradictory, showing again how fictitious capitalist conservation is becoming.

Unfortunately, it is difficult to see this contradiction for what it is, which leads me to a second reason why it is important to emphasize circulation. This is because of Marx's ([1867] 1976, 253, 381, and see above) argument that circulation develops into a "coercive external force" that becomes "an end in itself." Of course, production, distribution, and consumption can also become "ends in themselves," yet it is only their converging totality aimed at accumulation *through circulation* that becomes a "coercive external force." Hence, while circulation itself is indeed (continuously) produced, distributed, and consumed, as a totality it seems to have become an external force that affects us all—albeit in highly differentiated ways.

This, in turn, is further intensified in the context of hypercapitalist circulation, a maelstrom that moves at incredible speed and velocity, continuously taking on (and shedding) bodies, information, technologies, natures, relations, spaces, and time as it proceeds. Hypercapitalism, as Graham (2007, 1) stresses, is "hyper" indeed, creating the possibility that its circulation has significant potential to be used and abused as a seemingly external force that magically creates value for those who can step in and out of this circulation when they want to (see also Marazzi 2011). While we see the evidence of this all around us, particularly in the financial sector, we again immediately have to stress the limits of circulation as an "external force," since the growth of the circulatory circuit of production, distribution, and consumption of capital and values can absorb only so many "free riders." In other words, somewhere, someone still has to produce, distribute, or consume something, or, paraphrasing Leyshon and Thrift (2007), speculative structures can only be built on more mundane structures, and these are interwoven in complex ways.[10] Similarly, fictitious conservation has its limits and is thus never truly free from more traditional forms, even though these limits are always continuously pushed under capitalism.

Concluding Thoughts

Conservation, it seems, is increasingly becoming its own negation. Where once it might have been a Polanyian countermovement against the ecological contradictions of capitalism, this is no longer the case (Igoe, Neves, and Brockington 2010). Capitalist conservation has become an important instrument for the production of surplus value on its own and a way to "offset" and so seemingly legitimate more conventional methods of producing capital. This has meant that conserved nature itself needed to become capital, to become "value," and to be able to circulate within the ephemeral hyperspheres of contemporary capitalism. Marx ([1867] 1976, 638), while recognizing that the soil was one of the "original sources of all wealth," believed that capitalist commodities could only ever have value if they incorporated the interaction between labor and material nature (see also Arendt [1958] 1998). These days, we see something different. Humanity has become so fearful of its own capability of destroying all this wealth that it is increasingly "willing to pay" for its value to be recognized on the explicit condition that it does not incorporate the interaction between labor and material nature.[11] Characteristically, it does so by further bringing inherent contradictions in capitalism to new heights and levels, in this case to what I have called "fictitious conservation."

Fictitious conservation precariously tries to link the conservation of material nature via its "environmental services" to contemporary hypercapitalism and its emphasis on the circulation of ephemeral values. Occasionally it might succeed in doing so and indeed "save" some material nature from the onslaught of more "traditional" capitalist expansion. This, however, cannot be concretely verified if, for all intents and purposes, the link between consumers of liquid nature and the conditions and relationships that produced it has been severed. But this is hardly the point. The central paradox of fictitious conservation is not that it has little chance of not "working" but rather that it ultimately is not really about conservation at all. It is first and foremost about capital: generating value that is of use in and to contemporary capitalism. This is, I argue, what the severing of the link between material natures and ephemeral values signifies. Ironically, conservation's latest financialized products, such as conservation derivatives, "sustainability enhanced investments," mitigation banking, biodiversity offsets, and others, are still "marketed" under the heading of "environmental services" to try and emphasize direct links with material, biophysical natures. But it is the attempt at delinking that made these schemes attractive to capitalists, and this should therefore be the starting point of their characterization.

If this sounds "cynical," I would argue that it is—unfortunately—only the start. Truly cynical is that it no longer matters that in the complexity of turning conserved nature into capital, conservation has become fictitious; it can still sell. All that it needs is a compelling brand: a memorable logo, some catchy slogans, smooth marketing campaigns, visually captivating websites, celebrity spokespeople, and a take-home message that "everybody wins." It can make people "feel good" in the face of serious problems that seem to be going out of the rational, technical control capitalism thrives on. No wonder, then, that Sian Sullivan (2009, and see her chapter in this volume) talks about a profound manifestation of "cultural poverty" through the seeming incapacity to think of nature as anything in any other but capitalist terms.

Yet none of this is unforetold. Fictitious conservation is but one manifestation of the *intensification* of capitalism rather than its extensification (Smith 2007), and in line with Carolan's critical realist distinction between Nature, nature and "nature," the point for capitalist expansion is to penetrate *deeper* into rather than merely wider across reality. Hence, the uptake of conservation into the capitalist system signals that the hegemony of neoliberal capitalism is strong indeed, despite or perhaps even because of the recent crisis (Igoe, Neves, and Brockington 2010; Büscher and Arsel 2012). Indeed, the incorporation and celebration of its own contradictions may well be the basis of our current hegemony's perhaps unprecedented strength. To believe that nature can be conserved by increasing the intensity, reach, and depth of capital circulation is arguably one of the biggest contradictions of our times. The only way, then, to confront the contemporary contradictions around conservation is by working from and acknowledging both the "deep structures" and the contemporary dynamics of capitalism lest we continue to have conservation politics and policies based on symptoms rather than real causes.

Notes

1. See http://www.unep.org/pdf/A_Global_Green_New_Deal_Policy_Brief.pdf, p. 4, accessed September 15, 2010.

2. Neil Smith (2007) has written an extremely interesting and relevant essay entitled "Nature as Accumulation Strategy" that touches on many of the issues discussed in this chapter. In my view, however, Smith does not give "conservation" a central enough place (indeed, he hardly even uses the concept at all) and so misses some crucial links in explaining "conserved nature as capital" and what this implies for the value of nature in contemporary capitalism. These will be discussed later in the chapter.

3. Note that this is not the same as Marx's base-superstructure theory in relation to capitalism.

4. The category of "cultural ecosystem services" is interesting in relation to Sullivan's (2009) point that the whole exercise of subjecting nature to capitalist market dynamics is a profound manifestation of "cultural poverty," as it almost seems to acknowledge this very point by ensuring that "culture" is given its appropriate place in an otherwise culturally lifeless framework.

5. Importantly, Carolan (2005, 401) adds that "all three natures—'nature,' nature, and Nature—represent bounded hybrids. In each, sociobiophysical interactions occur, but to various degrees, thereby underlying the need to conceptually stratify reality so as to better understand how those strata interact and the bounded hybrids that result."

6. For "entrepreneurs" making the market liquid, see http://www.ecosystemmarket place.com/pages/dynamic/article.page.php?page_id=7682&section=news_articles& eod=1, accessed September 21, 2010. Important to add is that the degree to which this "rendering liquid" varies in practice is great.

7. Note that it is generally accepted that "more" liquidity is not always the best for market stability and thus for profits, hence the phrase "getting market liquidity right."

8. Although obviously not for all—many people in the world are still clearly caught in capitalist relations that are not all that hybrid, as conceptualized here.

9. Moreover, the term is actually confusing, as it seems that the "conservation mode of production" is somehow different from the "capitalist mode of production," while Garland (and others; e.g., see Brockington and Scholfield 2010) indeed argue that conservation is a capitalist mode of production and not a self-standing mode.

10. The simplistic way in which Mandel, Donlan, and Armstrong (2010, 49) argue that "short-term volatility in the price of the derivative does not affect the underlying asset" is therefore wholly unfounded and a disturbing act of wishful thinking.

11. Finding out people's "willingness to pay" for conservation is one of the favorite subjects of much mainstream ecological economics literature, as though this is synonymous with "legitimacy."

CHAPTER NINE

Nature on the Move II

Contemplation Becomes Speculation

Jim Igoe

In the first intallation to this triptych, Bram Büscher posits the emergence of a "liquid nature"—a kind of "fictitious capital" no longer grounded in any specific material context or relationships.

Abstraction and financialization "are extending new possibilities for nature's speculative release into the realm of circulating money" (Sullivan 2013b, 208). Liquid nature, Büscher (this volume) further argues, requires "'fictitious conservation,' conservation without any direct basis in material, sociobiophysical nature." Through fictitious conservation, the valorization of actual conservation activities is alienated from those activities themselves. Fictitious conservation circulates with liquid nature, which it also authenticates and valorizes. Both nature and the conservation of nature have been rendered into circulating commodity forms.

While these developments may initially appear as sudden and counterintuitive, emergent forms of nature for speculation are actually rooted in older and more widely recognized forms of nature for contemplation. Lukács (1971) has ascribed the pervasiveness of contemplation in modern society to Marx's concept of commodity fetishism, arguing that it is symptomatic of a generalized separation accompanying the alienation of labor's use value into exchange value by industrial capitalism. Over time, he asserts, people have increasingly become passive contemplators of the apparently autonomous movement of commodities as a "kind of second nature" (128).[1] The industrial transformation of commodity into a kind of nature was accompanied by a corollary transformation of nature into a kind of commodity, a spatially framed and putatively timeless view that people would pay to contemplate at a comfortable distance (Cronon 1996; Neumann 1998). This was consistently achieved by the forced removal of people who lived and labored in landscapes and the subsequent erasure of those removals (Igoe 2004).

Thus, as Sian Sullivan (this volume) elaborates in the third installation of this triptych, making nature move first required making it sit still as an increasingly deadened object of contemplation.[2] The second section of this essay will accordingly examine how the putative stillness of nature for contemplation has been entrained to the movement of nature for speculation. My analysis is informed by Guy Debord's ([1967] 1995) concept of "spectacle," a uniquely specialized and powerful form of "capital accumulated to the point that it becomes image" (thesis 34) and that mystifies and mediates the relationships of its own production (thesis 4). Debord further argued that spectacle's power to transform fragments of reality into a visually pervasive totality produced "a separate pseudo world" (thesis 2) offered in exchange for the totality of actual activities and relationships, a world of "money for contemplation only" (thesis 49).

Abstraction of nature into spectacle, as we shall see, has turned it into money for contemplation *and* speculation. Via multi-billion-dollar film and advertising industries, nature has moved onto screens that are seemingly everywhere (Mitman 1999; Brockington 2009). Such images also lend themselves to the simulation of nature in themed environments through which multiple and far-flung natures can be contemplated in one comfortable and conveniently located setting (Wilson 1993; Igoe 2004). Moving images of nature move consumers to buy products, take vacations, and give money to worthy conservation causes (Igoe 2010). Finally, spectacle provides visual testimony for a movable nature that can be "disassembled, recombined, and subjected to the disciplinary design of expert management" (Luke 1999, 142). This is the basis for what I call "ecofunctional nature," which appears as though it can be calibrated to optimize ecosytem health and economic growth. Ecofunctional nature, I will argue, is indispensable to the current global policy consensus that the financialization of nature is the key to its salvation—a pseudoqualitative accompaniment to complexly quantified forms of financialized liquid nature.

In addition to its abstraction of nature into circulating images and its visual embellishment of the practices and rationale of nature's financialization (cf. Debord [1967] 1995, thesis 15), spectacle offers a powerful technique for fostering and managing subjectivities appropriate to commodity nature (cf. MacDonald 2010a). The ability to create the appearance of certain realities, even when those realities have not been — in fact cannot be — achieved, is in itself a powerful effect, particularly when the reality in question is presented as "nature," "the inherent force that directs the world, human beings, or both" and "the material world itself" (Williams [1976] 1983, 219). Spectacle should therefore be considered as part of the wider mosaic that Michel Foucault (1982, 2007, 2008) called techniques and technologies of government (Debord [1988] 1998, 2; Crory 2002, 456).

In section 3 of this essay I will address the ways spectacle is produced and deployed in the intentionally modified and interconnected contexts that I call micropolitical milieus of commodity nature. These milieus are sites for the production and consumption of liquid nature and fictitious conservation, as well as of diversity of decentered and seemingly unrelated struggles over what nature is and what it will be. One of my main motivations for sketching these milieus is the possibility of short-circuiting spectacle and its attendant mystifications through the intensification of "channels, concepts, and processes that can link up and thereby intensify transversal struggles into larger collective, but discontinuous movements" (Nealon 2008, 106).

The Nature of Spectacle and the Spectacle of Nature

Historical treatments of nature, on the one hand, and spectacle, on the other, to the best of my knowledge have yet to be synthesized. The genealogical synthesis presented here focuses specifically on Western and predominantly North American contexts. I begin somewhat arbitrarily, with eighteenth-century land enclosures that accompanied Europe's industrial revolution and segregated countrysides into landscapes of production (for the production of wealth) and landscapes of consumption (for leisure and contemplation only) (for details, see Green 1990; Neumann 1998; Igoe 2004). The creation of nineteenth-century American parks as the ultimate landscapes of consumption revitalized this segregation to generate a widely recognized and eminently transportable abstract category of nature as big outdoors (Cronon 1996).

While such abstraction is an important element of circulating commodity forms, the category of nature itself was consistently presented as immutable, immovable, and thus forever outside of capitalist value production (Brockington, Duffy, and Igoe 2008). Landscapes of production, by contrast, were celebrated, elaborated, and simulated by elaborate commodity displays, mass-produced and embedded in new landscapes of consumption, from county fairs to world exhibitions. These displays not only effaced the labor that produced them but also appeared capable of transcending their own materiality (Connerton 2009), qualities that were important antecedents to what Debord would later call spectacle. Nature and spectacle thus appear less as separate parallel threads than as strands of a double helix becoming more tightly interwoven over time.

The Nature of Spectacle

As noted by Crory (2002, 457–58), Debord dated spectacle's origin to the year 1927 and "the technological perfection of the television. Right at the age when an awareness rose of the age of mechanical reproduction, a new model of circulation and transmission appeared. . . . [S]pectacle was to become inseparable from this new kind of image and its speed, ubiquity, and simultaneity." This year also introduced the first sync sound films, which demanded more concentrated attention from viewers than previous moving pictures. Debord's concern with sync sound suggests that he saw spectacular power as "inseparable from a larger organization of perceptual consumption" (Crory 2002, 458)—as near as possible to a total sensory experience.

Shortly thereafter, the Third Reich and Stalinism demonstrated the power of these technologies for producing encompassing state-sponsored propaganda, which Debord ([1988] 1998, 8) called "concentrated spectacle." American corporations and marketing firms deployed the same technology to produce "diffuse spectacle," an apparently decentered profusion of commodities on display (8). While doubtlessly catalyzed by these technologies, diffuse spectacle is rooted in mid-nineteenth-century world exhibitions that inspired German economists to posit an "exhibition value to indicate the productive capacity of representation itself. . . . [T]hings gain value simply by their mode of appearance, quite apart from their use value" (Brain 1993, 13–14).

Exhibition value proved and capitalized upon Marx's ([1867] 1976, 163) point that a commodity is "a very queer thing, abounding in metaphysical subtleties and theological niceties." By exaggerating and manipulating the metaphysics of commodities, their use value was effaced in what Benjamin (1979, 152) described as "a phantasmagoria that people enter to be amused." These were intentionally designed to overwhelm and disorient: giant glass buildings presented "an unending perspective that fades into the horizon," exhibit machines were also exhibiting machines, and panoramas moved past stationary spectators to simulate a hybrid collage of travel experiences (Brain 1993, 39, 48, 65). Such simulacra, Jameson (1991, 18) held, "come to life in a society where exchange value has been generalized to the point at which the memory of use value is effaced." By the turn of the twentieth century, exhibition simulacra were bursting their boundaries and spilling into their surrounding environs. Visitors to the 1900 Universal Exhibition found it difficult to distinguish the exhibition space from the rest of Paris (Brain 1993, 10). This, argues Connerton (2009, 60), was the beginning of diffuse spectacle, "an all embracing medium where people continuously interact with commodities."

Today this medium is indeed a kind of "second Nature," readily and ubiquitously visible in the environments with which consumers most commonly interact: cities, restaurants, freeways and rest stops, shopping malls, airports, train stations (and of course trains and planes themselves), not to mention theme parks and all manner of entertainment venues and tourist attractions.[3] It is also working its way into places like schools, hospitals, and office buildings. All these environments incorporate a diversity of video screens, from towering Jumbotrons to tiny televisions in taxicabs and airplane seats. They also provide settings for the production of commodified images, resulting in a recursive relationship between "reality and image" (e.g., a Jumbotron in Times Square promoted the Broadway production of *Madagascar* by endlessly repeating a sequence from the film in which the animals escape from the Central Park Zoo and wind up in Times Square). This is the basis of what Debord ([1988] 1998, 9) called "integrated spectacle—spectacle that has integrated itself into reality to the same extent that it was describing it, and that it was reconstructing it as it was describing it."

Since Debord's death in 1994, the boundary between actual and virtual reality has been further blurred by Wi-Fi and a diversity of portable communication devices. In my classroom a phalanx of glowing Macintosh logos mediates the space between me and the students, who are in actual and virtual reality at the same time. They listen to my live lecture and take notes while texting each other, shopping online, and updating their Facebook profiles. To spice things up, I show a YouTube video of Slavoj Žižek lecturing from *First as Tragedy, Then as Farce*, saving myself the labor of preparation and them the labor of reading. Žižek defines "cultural capitalism" as a reality in which "the very act of consumption entails redemption for being a consumer."[4] I rush to relate this point to the "prosumption" (simultaneous consumption and production) of nature (Büscher and Igoe 2013). I display a website where users can track radio-collared polar bears to see how drinking Coca-Cola helps protect Arctic habitats. Another invites consumers to adopt acres of virtual rain forest personalized with their names and a graphic of their favorite endangered species, but a synchronized closing of laptops indicates time is up. Next time, I promise, we will explore the transformations of nature that have rendered such presentations plausible.

The Spectacle of Nature

NATURE FOR CONTEMPLATION . . . . In contrast to nineteenth-century exhibitions, which enshrined intensifying industrial production, nineteenth-century national parks enshrined a special kind of "natural legacy." While these exhibitions offered escape from industrial life into phantasmagorias where "commodities are now all there is to see" (Debord [1967] 1995, thesis 45), parks offered escape from industrial life into putatively pristine realms, one of the main attractions of which was that commodities seemed to be absent (cf. Cronon 1996). In spite of these differences, exhibitions and parks operated by similar logics of abstraction and contemplation at play in the production of contemporary spectacle through which nature is now explicitly presented as the ultimate commodity.

Like exhibitions, parks effaced the conditions of their own production. Their displays of timeless wilderness for leisurely contemplation depended upon the systematic clearance of their human inhabitants.[5] For the illusion of timelessness to be effective, however, "this process of erasure had to erase itself" (Igoe 2004, 85). Nature was thereby presented as reality without social or historical connections, an arrangement ironically requiring significant administrative and technical intervention. The contemplation of nature in these terms, as Cronon (1996) aptly notes, was only possible by virtue of the modern conditions to which it was supposedly the antidote. For elites who championed American parks, however, this nature was nothing less than "the basis of universal truth available through direct experience and study. To study a particular instance offers a window onto the universal" (Tsing 2005, 97).

These conditions present four important antecedents to spectacle: (1) forgetting, (2) abstraction, (3) reifications, and (4) protoexchangeability. Forgetting is essential to Lukács's (1971) theoretical elaboration of commodity fetishism: "The precise process that produces commodities gets forgotten . . . [and] . . . manufactured artifacts . . . fall prey to cultural amnesia" (Connerton 2009, 43).[6] Forgetting is a precondition of reification, whereby artifacts appear to take on a life independent of their manufacture, "much like the laws of nature" (43). It also figures centrally in abstraction, whereby artifacts apparently transcend their own material limitations (Büscher this volume). The notion that individual parks materially embody an ideal universal nature is a kind of abstraction, since this universal nature presumably transcends the material boundaries of any park in particular. The ability of one object (a park) to stand for a class of objects (imagined universal nature) is furthermore an essential element of Marx's ([1870] 1978) theory of how commodities gain exchangeability and the basis of spectacle as "money for contemplation."

Because parks were meant to be stable and enduring, however, the nature they displayed required further mediation to transgress its material boundaries. This came with the aforementioned advent of sync sound and television, paving the way for the nature film industry. By the 1950s technicolor nature films were a popular staple for Western theatergoers (Mitman 1999), while freeways in the United States were transforming parks from a rarefied elite playground into popular vacation destinations for millions of newly affluent automobile owners.[7] Nature became part of the wider current of consumptive experiences that exploded on the scene in the years following World War II (for details, see Wilson 1993), presenting unprecedented possibility for its refinement into reified commodity forms that are also generators of additional value.

The career of Frankfurt Zoological Society director Bernhard Grzimek poignantly illustrates these refinements. At the end of World War II, Grzimek set up shop in what would become Tanzania's Serengeti National Park. With revenue from his award-winning film, *No Room for Wild Animals*, he conducted an aerial survey of the now world-famous wildebeest migrations. The survey was the centerpiece of *Serengeti Shall Not Die!*, an international best seller that won the Oscar for best documentary in 1957 (Bonner 1993). By the 1960s Grzimek presented a popular German television show called *A Place for Animals*, which he used to market nonexistent tours to East Africa. He speculated that this would generate sufficient demand to bring the safaris into existence, and he was correct (Lekan 2011, 225). Tourism is now Tanzania's second largest source of foreign currency (Igoe and Croucher 2007), while the royalties from *Serengeti Shall Not Die!* have built a world-class headquarters for the Frankfurt Zoological Society inside Serengeti, where it remains to this day (Bonner 1993).

Grizmek's story reveals nascent formulations of a now fully blown "conservationist mode of production" in which, "through various mediations . . . natural capital is converted into capital of a more circulating and globally ramifying kind" (Garland 2008, 62). This is achieved in large part through the abstraction of nature into image. In addition to their multi-billion-dollar value in the nature film industry, these images inform completely fabricated pseudonature in 3D blockbusters like *Avatar* and *The Lorax*. Images

212 • Jim Igoe

of conserved nature, and promises of conserving nature, are used to market everything from fast food to dish soap, SUVs to computer printers. They spread through the theming of space in airports, resorts, shopping malls, zoos, botanic gardens, and of course theme parks (Igoe 2010). Finally, as we have just seen, conservation NGOs use them to distinguish their brand in a crowded and highly competitive funding environment (Sachedina 2008). When images of nature are deposited in "image banks" (Goldman and Papson 2011, 137) from which they can be withdrawn and reanimated for any of the purposes above, there can be no further doubt that nature is "money for contemplation."

BECOMES NATURE FOR SPECULATION . . . . But how might nature that is money for contemplation become nature that is money for speculation? Both require abstraction and reification, but in the case of the latter, these are more meticulous and precise. As recent work by Sullivan (2013a, 82) illustrates, the abstraction of nature into tradable units of financial value is closely associated with "[V]ariously marketized forms of environmental offsetting," which will reputedly resolve "contradictions between economic development and nature health." Monetized ecosystem services theoretically correspond to land-based localities, nature banks, "where they can be situated and accounted for" (Sullivan 2013a, 83). These notional connections inform "key design features" for turning nature into money for speculation (Sullivan 2013a, 83; see also Büscher, this volume; Fairhead, Leach, and Scoones 2012b).

Two of these are of particular relevance to the present discussion. The first is the need for an "ecosystem metrics to permit exchangeability," a "symbolic numerical signifier that can serve as an abstraction of ecosystem aspects in different places and in different times, such that these abstractions become commensurable with and substitutable for one another" (Sullivan 2013a, 84–85). The second is the principle of "additionality," which assumes that nature conservation would not have occurred without offset payments (86). While the illogic of these assumptions may seem self-evident, it merits brief mention here: making nature quantitatively fungible conceptually obliterates the unique qualities of specific ecosystems and the cultures of people who dwell within them, while the principle of additionality depends on counterfactual scenarios.

It is precisely in areas like this that nature for contemplation is most important to nature for speculation. The former does not become the latter by turning into it; instead, it is like a becoming outfit that enhances someone's attractiveness to the point of becoming indistinguishable from them (as when we tell a friend, "That outfit is you!"). Nature for contemplation suits nature for speculation, covering over its blemishes and lumpy bits while enhancing its finer qualities. Nature for contemplation is "the indispensable decoration" of nature for speculation and "the general gloss on the rationale of the system" that produces it (Debord [1967] 1995, thesis 15).[8]

Productions of nature for contemplation have consistently and elaborately effaced its use values and its wider ecological and social connections (Cronon 1996). Contemplative activities are accordingly portrayed as nonconsumptive and transcendent of more mundane concerns, such as environmental effects of the contemplator's everyday activities (Cronon 1996) or even of traveling to the nature that will be contemplated (Carrier and Macleod 2005). The production of nature films and related conservation celebrity contributed to a popular perception that such natures would disappear if not for the efforts of heroic conservationists (Bonner 1993; Brockington 2009; Lekan 2011). Finally, mass-produced images and simulations of nature replaced uniquely contextualized qualities with iconic signifiers that could be transported to other locations and rearranged as desired (Wilson 1993). In this light, nature for contemplation appears tailor-made for scenarios of exchangeability and additionality, and it also becomes the idea that local people will prosper more from nature's exchange values than from its use values.

Considering these compatibilities, it is not surprising that nature for contemplation is a consistent backdrop to the reified practices that Büscher (this volume) calls "fictitious conservation" as well as standing for its putative ends. Fictitious conservation, Büscher correctly notes, is indispensable to the valorization of nature as money for speculation, which he calls "liquid nature." It is visually articulated—and made to circulate—by spectacular presentations of conservationists in action, often also incorporating narrative testimonies from conservationists themselves or celebrities speaking on their behalf (cf. Brockington 2009; Igoe 2010).

Nature for contemplation also figures in the calculative and technical reworkings of nature into money for speculation. The web page of TEEB (The Economics of Ecosystems and Biodiversity) tells us, "You cannot manage what you do not measure."[9] Of course, most people find it difficult to relate to abstract calculations and financial mechanisms, and nature for contemplation therefore remains essentially important. The TEEB page accordingly features a montage of endangered species, stock market trading screens, pristine landscapes, bar charts, and local people. A video promoting ARIES (Artificial Intelligence for Ecosystem Services) intersperses images of wildlife and satellite maps with illustrated explanations of how the technology operates to calculate values of environmental assets.[10]

Visual mashups of nature for contemplation, fictitious conservation, satellite maps, graphs, and charts are transforming nature for contemplation through explicit, though selective, presentations of what has long been present but previously hidden from view, "the application of techniques, procedures, and practices" by which nature is brought forth as "an object of knowledge and target for regulation" (Bäckstrand 2004, 703; cf. Foucault 2007, 79). Through the rapid proliferation of these kinds of mashups, even in popular presentations, nature for contemplation appears increasingly ecofunctional, still beautiful and entertaining but no longer pristine and best left to its own devices. Ecofunctional nature, as I call it, appears amenable to technological reorderings that will optimize economy and ecology, or at least accommodate putatively inevitable growth with minimal disruption to ecosystems and human well-being.

Popular presentations of ecofunction appear to operationalize cultural capitalism's promise of consumption redeeming consumption (see note 3). Donations and purchases appear to initiate events resulting in the protection of animals and ecosystems (Brockington, Duffy, and Igoe 2008, chap. 9; Igoe 2010). Texting "tree" to a designated number helps to make a shimmering virtual forest grow on Jumbotrons in Times Square, metaphorically standing for actual forests being planted in Kenya and Mexico.[11] Those who want more details of how such arrangements work can track virtual polar bears, follow the blogs of African conservationists, or watch videos outlining the logic of interventions they are helping to support (Brockington, Duffy, and Igoe 2008, chap. 9; Igoe 2010).[12]

Ecofunction also informs more general commentary on the environment in popular media. A recent special edition of *Time Magazine* (March 12, 2012), for instance, showcases a top 10 list of "ideas that are changing your life"—number 9: "Nature Is Over." The corresponding article (Walsh 2012) explains that we are living in what atmospheric chemist Paul Crutzen calls the "Anthropocene," a geological epoch in which human activity has become an irreducible element of the biological, chemical, and geological processes of our planet: "It is no longer us against nature," Crutzen opines, "instead it is we who decide what nature is and what it will be" (Walsh 2012, 84). This, the article continues, will revolve around technological interventions and their acceptable trade-offs. With genetically modified seeds we will grow more food on less land, freeing up space for wildlife. We will also "learn to live" with nuclear power's "risk of accident." Finally, we may have to "fiddle with the climate" using "planetary scale technology" (85).

While such scenarios are scary, they are made to seem less so by more whimsical interactions with ecofunction and language that lionizes the power of expert knowledge while softening the potential dangers of the transformations experts will oversee. While optimal ecofunction is almost certainly unachievable, in spectacle it can be conjured as a fait accompli. Spectacle's ability to project unity and consensus where none actually exist (Debord [1988] 1998, 2) makes it a powerful "technology of government" (cf. MacDonald 2010a). It provides visual articulations of nature as an ecofunctional object of intervention while concealing and marginalizing alternatives and opposition to its seemingly monolithic vision. We now turn from the relationship of spectacle to what I call the micropolitical milieus of commodity nature.

The Micropolitical Milieus of Commodity Nature

Spectacular celebrations of fictitious conservation and financialized nature conceal a much more contested politics of what nature is and what it will be. Missing are Western conservationists who believe in their bones that capitalism and profit motive spell nature's demise and not its salvation (see esp. Ehrenfeld 2009). We will also never see the occassional tourists who look beyond the spectacle they have been shown to gain a more nuanced understanding of nature conservation in specific locales. Some tourists even go to the trouble to educate others by disseminating what they have learned.[13] Also absent are the resistances and critiques of the diverse rural people whose lives, livelihoods, and ontologies of more-than-human reality have been discounted and displaced by conservation (see Dowie 2009; regarding ontologies, see Sullivan 2009).

In stark contrast to earlier green Marxist predictions that a looming environmental crisis would catalyze mass social movements, demanding ecologically sane alternatives to capitalism (esp. O'Connor 1988), the struggles of these actors are decentered and seemingly disconnected. My theoretical framing of these struggles draws from the productive intersection of Marxian concerns with the subsumption of culture by capital and Foucaultian scholarship on techniques of government. The conditions described in the previous sections reveal what Nealon (2008, 84) describes as the recirculation of value not only at all points on the socius but also at diverse points of interaction between humans and more-than-human nature around the world. Furthermore, as Read (2003, 126) has argued, the spread of commodity relationships from concentrated sites of production has required a concomitant spread of techniques and technologies designed to produce appropriate subjectivities. Nature on the move, which is produced and supported by these dispersions, presents a difficult moving target for activists and social movements, shifting and changing at different scales and locales.

This situation reflects two broader historical transformations that I have already touched upon. The first began when the nineteenth-century crisis of capitalist overproduction prompted the creation of a marketing industry to channel human desire into an apparently unlimited demand for consumer goods and services (see Debord [1988] 1998, thesis 45). The second began with the late twentieth-century proliferations of fictitious capital "without any material basis in commodities or productive activity" (Harvey 2006a, 95), of which reified nature for speculation is the most recent expression. Taken together, as they frequently are, these processes have spawned a gigantic intellectual labor force tasked with creating, celebrating, authenticating, and valorizing the latest consumer commodities and financial products. And of course there is the labor of consumption, which includes interpreting—and ideally taking appropriate action upon—a continuous bombardment of commodity signs: brands, slogans, and associations between desired experiences/qualities and designated products/ services (Goldman 1994).

All of this "immaterial labor," according to Read (2003, 129–30), both targets and shapes social communication and social space. It travels through "epistemic, aesthetic, and affective models that structure social communication." These, according to Virno (1996, 23), include "information systems, epistemological paradigms, and images of the world" and are communicated through manuals and reports, videos, seminars, and workshops. They are thus stored in archives but also in the "minds of workers, as little productive machines (virtual fixed capital), without necessarily originating from them" (Read 1993, 131). These valuable little machines are activiated and reproduced in realms outside the direct control of capital: in the subjectivity of producer/consumers and the diversity of social spaces they inhabit.

In *Foucault beyond Foucault* (2008), Nealon describes how mutations in modes of production from factory to cultural life correspond to similar mutations in modes of power. My understanding of these mutations is informed by Foucault's (1982, 220) basic definition of government as the "conduct of conduct," achieved by "structuring the possible field of action of others. . . . [I]t induces, it seduces, it makes easier or more difficult." Government is inseparable from regimes of truth (Dreyfus and Rabinow 1983), producing objects of knowledge and intervention (e.g., conservation as a regime of truth that produces nature). Government is concerned with shaping people's subjective perceptions of what is possible, plausible, and desirable and thus of their own efficacy in any situation—for the purpose of "developing, canalizing, and harnessing social and individual capacities on a . . . cost effective mass scale" (Nealon 2008, 27).

Over time, Nealon (2008, 31) argues, techniques and technologies of government have become more efficient as they have been made lighter and more virtual. Discipline, for instance, works in a retail fashion on individual bodies in specific institutional contexts through a "series of discontinuous institutional training exercises" (41). Subsequent modes of biopower do not replace discipline but infiltrate it and amplify its effect by working throughout populations and infusing "each individual at a nearly ubiquitous number of actual and virtual sites." Biopower works less on actual bodies and more on potential actions, thereby "gaining an intensified hold on what [bodies] are, will be, may be" (31). Along these lines, Foucault (2008, 271) posited that neoliberalism is a new "art of government . . . which will systematically act on an environment and modify its variables." The point is to channel the acts of individuals, presumably acting in their own best interests, toward a spectrum of preferred outcomes and effects (Fletcher 2010).

What forms might "enviromental governmentality" take with respect to the politics with which we are currently concerned? The politics of commodity nature, I believe, occur for the most part in modified environments that greatly resemble Foucault's (2007, 20–21) discussion of milieu, which is a "multi-valent and transformable framework" fabricated from "pre-existing material givens" that are designed to "maximize the positive elements . . . [while] minimizing what is risky and inconvenient" (these elements, of course, are defined for the most part by planners, politicians, and other powerful actors). "It is what is needed to account for the action of one body on another at a distance." "What one tries to reach through this milieu is precisely the conjunction of a series of events produced by [people] and quasi natural events which occur around them." While his discussion is derived from town planning in eighteenth-century Europe, the dynamics he describes are visible, intensified, and refined in the micro-political milieus of commodified nature.

The first of these is a consumer milieu consisting of the kinds of spectacle-dominated environments described in the previous section of this essay. In this milieu the action of one body (a consumer) can appear to initiate a chain of events positively affecting another body at a distance (e.g., a polar bear or a tree). Its recent explosion of Web 2.0 applications marries self-expression (sharing your favorite causes) to wholesale monitoring and delineation of consumer types (people who care about the same causes as you also love

Endangered Species Chocolate!).[14] While micropolitics of commodity nature occupy a tiny segment of this milieu, its presentations of conservation and nature are dominated by celebrity, consumerism, and depoliticized presentations of fictitious conservation (Igoe 2010). While it is possible to find virtual communities and media that are critical of commodity nature, they are few, and their connections to efficacious action are undeveloped.[15] This remains for the most part a spectator milieu.

Next we have a transnational institutional milieu that corresponds to what MacDonald (2010a) calls "the new fields of conservation." This is the policy environment in which the creation and valorization of new forms of nature for speculation take place. It is also a realm in which immaterial labor takes the form of "little productive machines," like TEEB and ARIES, as described above, and many other formulas, models, and matrices disseminated through interactive displays, expert presentations, promotional literature, videos, seminars, workshops, and the like. Earlier in the millennium this milieu was more prone to conflict and contestation. The 2003 World Parks Congress in Durban, for instance, was disrupted by protests from indigenous peoples (Brosius 2004; Brockington and Igoe 2006). Similar disruptions have been reduced at subsequent events through a variety of management techniques designed to minimize interactions between attendees likely to have strong disagreements (MacDonald 2010a). They also entail orchestrated performances of community consensus, miniature concentrated spectacles hailing appropriate subjectivities in their intended audiences (MacDonald 2010a). This milieu, itself accessible to only a limited range of actors, is segregated into exclusive events within events that are accessible to only the most powerful and privileged actors of all.

Finally, we have the landscapes and seascapes that are sites to conservation interventions and the source of nature spectacle circulating for contemplation and speculation in the milieus outlined above (see Igoe 2010). The modification of these milieus increasingly turns on complex and multifaceted arrangements among NGOs, states, corporations, and local people operating through "the restructuring of rules and authority over the access, use, and management of resources, in related labor relations, and in human-ecological relationships" (Fairhead, Leach, and Scoones 2012b, 239). While these arrangements include voluntary relocation guidelines, they also often involve arrangements in which choices for relocation and/or livelihood transformations appear preferable to contending with the risks that the inverventions themselves present for existing settlements and livelihoods (see Schmidt-Soltau and Brockington 2007). Resistances to such transformations are complexly intertwined with "local cultural politics, identities and material struggles" and frequently informed by complex assessments of the situations in question. However, established presentations of local people as "green primitives" make it only too easy to reimagine these resistances as uninformed, "primitivist and hopelessly romantic" (Fairhead, Leach, and Scoones 2012b, 253).

Concluding Remarks

The micropolitical milieus that I have sketched here are currently the subject of intense scholarly analysis, and important inroads are being made into understanding their internal dynamics, their interconnections, and their disconnections. I hope that in some small way the conceptual schema I have offered in this chapter will prove useful to ongoing and future endeavors. If, however, "the point is to change it," there remain a few things to say. It almost goes without saying that the stakes are very high by just about any standard. As Sullivan argues in the third installation to this triptych, productions of nature for speculation are profoundly antiecological. Indeed, she puts it more strongly than this: they are made possible by the systematic deadening of animate ecologies and noncapitalist human ontologies. Nor is it likely that turning nature into a giant bundle of capital assets will automatically result in the global spread of holistic stewardship practices. To quote Fairhead, Leach, and Scoones (2012b, 244): "Logic might suggest that this would inevitably value ecosystems over and above the sum of [their] parts. And yet that is what employees often think of viable businesses they work for when they are sold—before they are asset-stripped. The perversities of the financialized world are legion, and once there are markets for nature's assets, so nature's assets can be stripped."

Debord ([1971] 2008, 81) perhaps put it most succinctly with his assertion that capitalism was creating "a sick planet" rendered palatable and seemingly inevitable by media spectacle as "the environment and backdrop of its own pathological growth and reproduction." As evidence for this undesirable outcome mounts, I increasingly hear conservationists lament that they did not know what they were helping to make when they embarked on the financialization of nature—a sentiment that resonates with Foucault's observation that "people know what they do; they often know why they do what they do; what they don't know is what they do does" (in Dreyfus and Rabinow 1983, 187).

Of course, it is doubly difficult to know what we do does from inside a spectacle-saturated milieu. As Agamben (1993, 80) notes in his comments on Debord's legacy, spectacle is "the alienation of language itself, of the very linguistic and communicative nature of humans." As such, Read (2003, 151) elaborates, "it is the simultaneous site of mystification and struggle." Spectacle, as a technique and symptom of power, works to appropriate the diversity and commonality of human communication and experience, presenting it as an apparently unassailable singularity. Spectacle's meta-message, Debord ([1967] 1995, thesis 12) believed, is "everything that appears is good; everything that is good will appear."

As both Debord and Foucault urged, each in his own way, we denizens of postindustrial consumer society have a lot of work to do on our sub-jective experiences of, and by extension engagements with, "the intense singularity that is the present" (to borrow a phrase from Nealon 2008, 106). More expansively, struggles in the micropolitics of commodity nature are animated by and productive of transformative knowledges and practices that need to be taken more seriously. To quote Foucault once more: "It is possible that [in] the struggles now underway, the local, the regional, [and we can add the transnational], discontinuous theories being elaborated in the course of these struggles, and which are absolutely of a piece with them, are just beginning to discover the ways in which power is exercised" (in Deleuze 2001, 212).

As West (2006, 66) aptly notes, for instance, we have not begun to understand the creative and diverse ways that people around the world engage in and critique capitalism and, by extension, capitalist natures. These in turn point to possibilities beyond oppositional critique, taken up by Sullivan in the following chapter, enlivening "both nonhuman natures and understandings of what it means to be human in intimate, moving and maintaining improvisations with other-than-human worlds."

Notes

1. This usage is distinct from current usages referencing anthropogenic environ-ments (Hughes 2005, 157–58), though all share Hegelian roots (see Schmidt 1971, 42–43; Smith 2008, 19; and Jappe 1999, 20–31).

2. The logic of deadened nature for contemplation is lucidly set out by Timothy Luke in his discussion of the Nature Conservancy as the Nature Cemetery. "Nature is dead," Luke (1999, 68) argues. "Material signs of its now dead substance need to be conserved as pristine preserved parts, like pressed leaves in a book, dried animal pelts in a drawer, or a loved one's mortal remains in a tomb."

3. For a detailed account of these transformations in North America, see Wilson (1993).

4. To view this video, visit http://www.youtube.com/watch?v=hpAMbpQ8J7g, accessed July 27, 2009. For the more adventurous there is of course the book by the same title (Žižek 2009).

5. For some time this aspect of parks was so underresearched that Jacoby (2001) described it as "the hidden history of American conservation" (also see Brockington and Igoe 2006). Since then the topic has gained more attention through a flurry of research, investigative journalism, and documentary films. For an overview of this extensive work I recommend Dowie (2009).

6. These ideas were a major inspiration for *Society of the Spectacle* (see note 1).

7. The enjoyment of pristine wilderness by millions of people was of course a paradoxical arrangement, as evidenced by "bear jams," which happen when the supply of bears cannot meet the demand of photographers, resulting in hundreds of tourists concentrating around sparsely distributed animals. Parks in Tanzania experience the similar phenomenon of "lion jams," and I imagine parks in India probably have "tiger jams."

8. An alternative translation is "insdispensable embelishment" [*sic*], http://www.bopsecrets.org/SI/debord/, accessed July 26, 2012.

9. TEEB is a global initiative and an evolving array of calculative technologies dedicated to saving nature through its systematic valuation. See http://www.teebweb.org/HomeofTEEB/tabid/924/Default.aspx, accessed July 26, 2012.

10. ARIES is a web-based technology offered to users worldwide to assist in rapid ecosystem service assessment and valuation. See http://www.ariesonline.org/about/intro.html, accessed July 26, 2012.

11. See http://3blmedia.com/theCSRfeed/Earth-Day-2011-Celebrations-Times-Square-ReGreen-World, accessed July 27, 2012.

12. See especially http://www.youtube.com/watch?v=_fwEwBdAM6U&feature=endscreen, https://www.arctichome.com/web/index.html, and http://www.youtube.com/watch?v=ACHqdkfmP4Q, all accessed July 27, 2012.

13. See especially "View from the Termite Mound" by Susanna Nordlund, http://termitemoundview.blogspot.com/, accessed July 27, 2012.

14. The Facebook page of Endangered Species Chocolate currently features a photograph of four lion cubs. Clicking on this takes you to a comment from a "friend," who states, "I officially want to adopt the four babies pictured here ♥ ♥ ♥ ♥ I know it's not reasonable, but they're just so stinkin cute!!!!!" The company responds, "We know the feeling! You can symbolically adopt them through African Wildlife Foundation." See http://www.facebook.com/EndangeredSpeciesChocolate and https://www.chocolatebar.com/categories.php?category=Gift-Collections percent2FAWF-Adoption-Collections, both accessed July 27, 2012.

15. For an example of a critical virtual community, see the Facebook page of Just Conservation, http://www.facebook.com/JustConservation, accessed July 15, 2012. For critical media, see *Silence of the Pandas: What the WWF Isn't Saying*, http://www.youtube.com/watch?v=YSztqfLT3F0; *Conservation's Dirty Secrets*, http://www.youtube.com/watch?v=pTVELt-pdGc; *A Place without People*, http://www.youtube.com/watch?v=QrEmUjNhwyo; and the BBC's *Unnatural Histories*, http://www.bbc.co.uk/programmes/b011wd41, all accessed July 14, 2012.

CHAPTER TEN

Nature on the Move III

(Re)countenancing an Animate Nature

Sian Sullivan

> From the invisible atom to the celestial body lost in space, everything is
> movement. . . . It is the most apparent characteristic of life.
> —ÉTIENNE-JULES MAREY, 1830–1904,
> cited in E. Oberzaucher and K. Grammer,
> "Everything Is Movement," emphasis added

> In the Oedipal relation the mother is also the earth, and incest is an
> infinite renaissance.
> —G. DELEUZE AND F. GUATTARI, *Anti-Oedipus*

> They lived firmly and wholly in the real world. Spiritual yearning and
> the sense of sacredness they knew, but they did not know anything holier
> than the world, and they did not seek a power greater than nature.
> —URSULA K. LE GUIN, *The Telling*

> To refract . . . *To change direction as a result of entering a different
> medium.* . . . *To cause . . . to change direction as a result of entering a
> different medium.*
> —from WIKTIONARY, the free dictionary

Coding Nature?

In the beginning, the primal Mother Tiamat was creator of the universe,
heaven and earth, water, air, and plants. This female serpent emerged
from the sea to teach humankind the arts of living well. Over time, a com-
plexified pantheon of gods and goddesses began to bear a curious resemblance to the egoic and heroic struggles of an emerging metropolitan elite.

But their transcendent and celebrated glamour was not without challenge. Threat came from their own laborers and from envious neighbors and other barbarians, not to mention the capricious dance of the elements, which brought drought, flood, and all manner of earthly chaos to challenge their elite order and control. It was clear to the ruling class that Tiamat needed disciplining. And so a young god killed Apsu, her favored consort, and then crowned himself king. With his wife, Damkina, he had a son named Marduk. This son was a murderer who was driven to crush all chthonic, chaotic threat to the growing Babylonian hierarchical order. He killed the genetrix Tiamat and from her split and dead body remade heaven and earth. From the blood of her murdered consort Kingu Marduk made humans to be slaves to the ruling-class gods, assistants in the latter's pursuit of war, leisure, and pleasure. This complete revolution *turned the generative cosmos into dead matter* to be fashioned for use through the artisanal expertise and force of the ruling class. Standing astride the dead body of the genetrix, the gods assumed transcendence over and possession of their new objects of the cosmos. The rest, as they say, is history.[1]

Bram Büscher and James Igoe, in the first two chapters of this "triptych," diagnose the contemporary moment as saturated with a dizzying range of commodified, financialized, and spectacularized "other-than-human natures."[2] Many of these are new commodities designed to service a green-economy suturing of economic growth and environmental sustainability (see UNEP 2011). This in part relies on market logic to solve the environmental harm caused through the failure of capitalist markets to adequately account for the costs of environmental degradation.[3] Carbon credits, environmental options and futures, biodiversity derivatives, mitigation insurance, species credits, biodiversity offsets, and so on are among the plethora of actual and proposed entities populating the resultant new ecology of monetized and marketized nature (Sullivan 2012b, 2013b). They are made through particular abstractions, significations, and conceptual transformations of nonhuman nature to create a circulating commensurability of environmental health and harm that can be managed through the remote control of the market. And they become visible through lively marketized exchanges in which the "value" of nature, as the dollar signs and zeros and ones of digitized "natural capital," becomes materialized (as described and discussed in Robertson 2006, 2012; Sullivan 2010b, 2012b, 2013a; Szersynski 2010; Pawliczek and Sullivan 2011; Bracking 2012; Lohmann 2012; also see Plant 1998).

At the same time, these universalizing abstractions seem to amplify and even require a deadening of nature's immanent and vivacious movement.

As Igoe (this volume) writes, "Making nature move first required making it sit still as an increasingly deadened object of contemplation."[4] The liquid, capitalized nature of which Büscher (this volume) speaks thus is simultaneously an abstracted, contemplated, and stilled nature legible to the extent that it can be packaged into units that can be calculated and traded, for "it is only when 'nature' is dead that a full-scale Nature™ Inc. becomes a possibility" (Arsel and Büscher 2012, 62). The commodity fetishism that animates capitalist circulation thus not only strips away the "incorporated creative life [of workers] toward equivalence within an exchange," such that labor value is deflected toward "the account of capital" (Nancy 2001, 3).[5] In the biopolitical subsumption of life itself, the "zombie-soul" (Holert 2012, 4) "animating" the commodity form also makes productively exchangeable but deadened objects of life's immanent vitality and diversity. The current reframing of a working nature as provider of discrete services (see Daily and Ellison 2002, 5) and as a bank of units of natural capital might thus be seen as an extension of "thanato-politics" and "necro-capitalism" (Banerjee 2008) in the environmental sphere, even while claiming exactly the opposite.[6] Through these new myths of nature (Sullivan 2013c), "the soul of capital" is able to extend its vampiric subjugation of life in the juggernaut momentum of "value" production, economic growth, and corporate power (Crouch 2011).

Current socioecological accounting practices conceived as emphasizing the monetized "value" of nonhuman nature (Costanza et al. 1997; Sukhdev 2010), such as in ecosystem service science, carbon metrics, biodiversity offset metrics, "the TEEB approach," REDD+ calculations and corporate ecosystem valuation (see, e.g., BBOP 2009, 2012; TEEB 2010; WBCSD 2011; DEFRA 2012), thus are conceptualizing and constructing other-than-human natures such that they can be further entwined and entrained with transcendent monetary categories and measures (see MacKenzie and Millo 2003).[7] These accounting practices attach monetary value to selected indices of nonhuman nature, and, notwithstanding the work of those in the CBD process to mobilize finance through enhancing regulatory mechanisms and fiscal reform, they are permitting the emergence of new market exchanges in these measures.[8] These practices perhaps do relatively little to respond to and transform the underlying value practices tending toward problematic nature exploitation and commodity fetishism (see Kosoy and Corbera 2010). Instead, they rely on economic incentives that appeal to individual self-interest so as to alter behavior. As such, these monetizations continue the zeitgeist of (neo)liberal individualism and competitive entrepreneurialism with which exploitative and dissociative socioenvironmental relations currently are linked. (For key proposals by significant corporate and financial "visionaries," see Kiernan 2009; Sandor 2012; Sukhdev 2012.) As MacDonald and Corson (2012, 159) claim, the "endeavour to put an economic value on ecosystems makes nature legible by abstracting it from social and ecological contexts and making it subject to, and productive of, new market devices." In a Foucaultian sense, new nature valuation technologies act to intensify capital's power effects (Nealon 2008; see also Sullivan 2013b), whereby all is subsumed to the "truth regime" and associated accumulations of "the market" (Foucault 2008). The subsequent release of new nature values into the totalizing and biopolitical control of the smooth flows of capital associated with globalized markets thus intensifies capital's power effects while also sustaining the subsuming dynamic of capital present since at least the European Enclosure Acts (Federici 2004).

In the process, new constitutions of material nature are brought forth, together with new means of its practical appropriation. The discursive and calculative technologies (see Callon and Muneisa 2005) that create and prime entities for marketized exchanges—from genetic plant resources under the United Nations Convention on Biological Diversity to insurance derivatives on the Chicago Board Options Exchange—thus structure and shape the materiality of the things that thereby become traded, with effects on the ecosocial contexts from which they derive (MacKenzie and Millo 2003; Brand and Görg 2008). At the same time, contemporary techno-configurations of circulating commodified nature are amplifying an ecology that resides in a radically disembedding and disembodying ontology. Through this, the fates of diverse rain forest assemblages are influenced and managed through the remote control of electronic exchanges;[9] online cybersafaris of African savannas seemingly generate authoritative knowledge of "the real thing";[10] and radical geographies of nonlocality become the basis of nature conservation through the marketized exchange of varied "conservation credits" between landowners and localities.[11]

These approaches to environmental management for conservation constitute both recent innovations and intensified conceptual decouplings of culture from nature familiar in Europe since at least the Enlightenment, an era that itself is rooted in Renaissance interpretations of classical Greek philosophy (Merchant [1980] 1989). They are part and parcel of a broader series of epistemic shifts that can be traced to successive transformational moments in different cultural milieus, such as that summarized in the Babylonian story with which this chapter opens (also see Merchant [1980] 1989; Roszak [1992] 2001). In the Western context, these approaches extend and entrench an older Occidental biblical creation hierarchy asserting "man's" dominion over other creatures (Cohen 1986, 15) and the dominion of a singular God over all. As returned to in the postscript to this chapter, the associated *transcendence* or "set-apartness" of experience of the sacred is a related and relevant construct flowing from this monotheism. It corresponds with both a removal of "the sacred" from the immanent vital materialities of "nature" and an associated separation of leader-priests from followers through variously rigidified hierarchies that serve(d) political, economic, and technological inequities (Young 2011).

The phenomena described above invoke a significant paradox: that of the intensified lively circulation of new commodified digital units of nonhuman nature intended to signify the incorporation of environmental harms into productions of economic value (what Büscher in this volume calls "liquid nature") and that of the simultaneous dependence of these lively representations and circulations on an amplified treatment of non-human nature as distant, stilled, bounded, and mute object (see Ingold 2006). A key effect of this, as Latour (2004) gestures toward in his *Politics of Nature*, is that human nature has been rendered increasingly deaf to a stilled and desacralized nonhuman nature that is its mirror (Weber [1904] 2001; Curry 2008). Environmental philosopher Andrew Dobson (2010) elaborates the implications of this, noting an associated entrenching of an Aristotelian position that "Man" alone is a political animal, with non-human nature rendered mute in political terms. Anselm Franke (2012b, 12–13, emphasis added) thus invokes Indonesian narratives that tell of "*the falling silent of the world* under the burden of 'primitive accumulation,' of capitalist exploitation, and of colonial administration." And so behind the contemporary proliferations and circulations of the fetishized abstractions of nonhuman nature described above is a deepened muting and deadening of the enunciative possibilities of nonhuman natures, accompanied by an intensified "tuning out," as irrelevant and obstructive, of the communiqués of other(ed) culturenature ontologies.

This predicament, and its tendency toward inequity and a possibly global ecocidal moment, generates significant questions. What relationships and ontologies are strengthened through these contemporary constructions and circulations? What is demoted and negated? And what "gaps" remain for (re)embodying socioecological arrangements that are both differently democratic and nourishing of life's alive diversity?

Deleuze and Guattari ([1972] 2004, 177–78), on whose work I draw throughout this chapter, refer to nature's immanence as "the germen," the original full and flowing body of the intense germinal and generative earth.

They argue that inhibition of the incest-like desire for possession of this full and flowing force has always required systemic cultural codifications. Thus, "in indigenous and other . . . rural communities of the world, one almost always finds institutions with rules that serve to limit short-term self-interest and promote long-term group interest" (Berkes 2008, 238). Indeed, for most of human history and cultural circumstances, the separating culture/nature assumptions described above seem to have been understood and refused as negative in their effects. As Deleuze and Guattari suggest, the abstracting and fictionalizing impetus that enables state capitalism's de- and recoding of the *ecosocius* has tended to be thoroughly resisted, prevented, and contained (cf. Clastres [1974] 1989; and see the discussion in Melitopoulos and Lazzarato 2012b). They write, for example, that "the primitive machine is not ignorant of exchange, commerce, and industry; it exorcises them, localizes them, cordons them off, encastes them . . . so that the flows of exchange and the flows of production do not manage to break the codes in favor of their abstract or fictional quantities" (Deleuze and Guattari [1972] 2004, 168; see also Polanyi 1944). Anthropologist Laura Rival (1996, 146) echoes this in writing of Huaoroni, Zaparo, Shuar, and Tukanoan nations of the Upper Marañón River of Peru that "they have constituted nomadic and autarkic enclaves fiercely refusing contact, trade, and exchange with powerful neighbours." As such, the separation of market exchanges from ecosocial relations (as in the ideal of free-market economics) has been variously inhibited in part because this separation is known to break embodied ties of living community, ties that otherwise might be understood as binding all emplaced entities in moral *and maintaining* economies of connection, cooperation, and sharing (Bird-David 1992; Lewis 2008/9; Graeber 2011).[12]

In the modern era of industrialism, capitalism, and the controlling freedom of the market, human endeavor has seemingly become untethered from these codes. The effect has been a chimerical disembedding of human from nonhuman natures (Polanyi 1944; Latour 1993) and an unleashing of accumulated stocks into flows that escape prior societal codifications (Buchanan and Thoburn 2008, 25). In this reading, it is an intensified breaking of inhibiting codes that makes possible current value-accumulating circulations of newly commodified stocks and flows of abstracted nature and whose recoding as "natural capital" and "ecosystem services" assists this instrumentalization perfectly.

I seek, then, to destabilize and refract these deadening and disembodying assumptions by calling on ethnographic and historical records that clarify different possibilities for conceptualizing and enacting human-with-nature existence. I focus on varied *animist* ontologies from different geographical and temporal contexts (see also Ingold 2006). Modernity's "nature as mute and stilled object" is an empowered but particular cultural fetish (or "factish"; cf. Latour 2010b), permitting instrumentalizing abstractions that are proving problematic in their socioecological effects (see Latour 2004; Hornborg 2006). Nature's conceptual pacification has been made possible precisely through denial and purification of the animist ontologies that both constitute modernity's necessary Other and pose(d) danger to the transcendent coherence of modern (b)orders (Franke 2012b; cf. Douglas 1966). As the Nobel Laureate and molecular biologist Jacques Monod wrote in the 1970s, science necessarily "subverts everyone of the mythical ontogenies upon which the animist tradition . . . has based morality" so as to establish "the objectivity principle" as the value that defines "objective knowledge itself" (1972, 160–64, quoted in Midgley [2004] 2011, 4). My intention instead is to refocus attention on the ecoethical effects that may be associated with bringing nature "back to life" via a reactivation of animist relational ontoepistemologies concerned with maintaining good relations between all entities/actants in each moment rather than conserving via capitalizing specific objectified and thus transcendent natures (cf. Harvey 2005; Ingold 2006; Bird-David and Naveh 2008; Sullivan 2010a; Curry 2011; Stengers 2012).

I hope to speak to Bruno Latour's (2010a) call, in his recent "Compositionist Manifesto," for movements beyond critique and toward curiosity and support for subversive everyday (re)compositions of human-with-nature ecologies. Latour encourages us to broach and brave, as well as to remember, a very different collection of concepts, concerns, and practices. In this vein, relevant work regarding diverse and (re)embodying insertions of nature and materiality in society is being productively conducted in a range of social science and humanities genres, including critical geography, science and technology studies, religious studies, feminism, environmental philosophy, political theory, and art (see, e.g., Castree and Braun 2001; Harvey 2005; Plumwood 2006; Curry 2008, 2011; Haraway 2008; Bennett 2010; Coole and Frost 2010; Lorimer 2010, 2012; Panelli 2010; Yusoff 2012; and the contributions in the volume of *e-flux* edited by Franke [2012a]). But the possibilities are greater still for "enlivening" nonhuman realms and the ecosocius in ways that refract the deadening abstractions of Nature required for its financialized circulations. Anthropology and cross-cultural ethnographic work can offer much here by way of bringing into the frame markedly differently embodied culturenature ontologies and associated effects (see Descola and Pálsson 1996a; Ingold 2000, 2006, 2011; Posey 2002; Hornborg 2006; Neves 2006, 2009b; Berkes 2008; Moeller 2010).

I explore a few contributions here, all of which have a key commonality. This is of an amodern assumption of the alive sentience of "other-than-human natures" as animate and relational subjects rather than inanimate and atomised objects. An effect is to enliven both nonhuman natures and understandings of what it means to be human in intimate, moving and maintaining improvisations with other-than-human worlds. "Animism" is the term used to describe this orientation. This is a descriptor that enfolds Edward Tylor's ([1871] 1913; see also Gilmore 1919) "mistaken primitives," positioned prior to the attainment of Enlightenment rationality in his theory of religion, with postmodern "ecopagans" of the industrial West, for whom animism is a contemporary ecoethical "concern with knowing how to behave appropriately towards persons, not all of whom are human" (Harvey 2005, xi; see also Plows 1998; Letcher 2003; Harris 2008). As such, animism is both "a knowledge construct of the West" (Garuba 2012, 7) and a universalising term acknowledging a "primacy of relationality" (cf. Bird-David 1999; Ingold 2006) and a set of affirmative practices that "resist objectification" by privileging an expansionary intersubjectivity (Franke 2012b, 4, 7). Animist ontoepistemologies in varied circumstances seem to have tended toward ordinary praxes of living with ecoethical effects that enhance(d) ecocultural diversity and poetic meaning. As such, they are worthy of (re)countenancing.

Counter-Culturenature Ontologies

> Countenance *n. 5 bearing or expression that offers approval or sanction : moral support. v. to extend approval or toleration to : sanction.*
> — MERRIAM-WEBSTER ONLINE DICTIONARY

An established ethnographic literature destabilizes some of the seemingly intractable dichotomies and categories infusing the growth- and commodity-oriented political economies of modernity and postmodernity. In this, the culture/nature dualism and accompanying assumptions of either environmental determinism (over cultural activity) or a passive Nature as background to cultural dominion make way for "ethnoepistemologies" that challenge these modern ways of organizing what it is possible to know (Descola and Pálsson 1996a; Hornborg 2006). Key here are a plethora of possibilities in which humans are envisaged as sharing ontological social space with the beings that "Western human ontology" (see Glynos 2012) frames as "nonhuman." This seems entwined with a sense that what exists is brought into being through ongoing participation in relationship by all entities (Ingold 2006). Agency, while differentiated, thus is present everywhere, such that all activity is simultaneously imbued with a moral, if frequently ambiguous, dimension (Ingold 2000). Arguably, such different culturenature ontologies have actualized lively embodied ecologies that favor the maintenance of biological and other diversities. As such, they warrant engagement and "re-animation" (Ingold 2006, 19) even in contexts more attuned to modern technological and economic discourses regarding policy solutions in biodiversity conservation (and perhaps especially in such contexts).[13] In what follows I draw on a selection of ethnographic studies to foreground elements of the animist socioecologies associated with several contemporary and historical circumstances. These emphasize what seems to be an uncynical ontology that knows all dimensions of existence to embody and enact agency in interrelationship, as well as to be animated and alive with sacred and connective meaning. I return to the latter theme in the postscript that completes this chapter.

My first exploration is a 1992 article by anthropologist Nurit Bird-David, whose ethnographic work on "animism" has been critical for establishing key parameters in this subfield. In this early article she develops Marshall Sahlins's (1974) conception of "the original affluent society" through considering so-called hunter-gatherer conceptions of the provisioning roles of other-than-human natures in such economies. Her ethnographies are of the Nayaka of South India (also see Bird-David and Naveh 2008), Batek of Malaysia, and Mbuti of Zaire. Their orientations to "nonhuman natures" are understood in terms of assuming "the environment" *to give to* humans in a profound "economy of sharing" that mediates human-with-human and nature-with-human provisioning. "Nonhuman" natures are "humanized" such that they are known as kin and as ancestral embodiments, as communicative agencies, and as friends. Landscape entities as well as nonhuman animal species are attributed with life and consciousness. An *order of goodness*, while at times ambivalent, in general is assumed. Such knowledges find expression in value practices oriented toward sung, spoken, and danced communication and multiway gift giving with nonhuman natures that are equivalently expressive. All of these situate human persons as agents continually doing their part to maintain a moral and dynamically generative socioecological order of trust that implicitly is assumed to be both abundant and good. This assumption of abundance and the associated "full-subject" (Glynos 2012, 2379) mitigates against a need for excessive consumption or hoarding of possessions.

Specific cultural innovations assist with the maintenance of this sense and assumption of abundance. Work by anthropologist Jerome Lewis (2008,

2008/9) with Mbendjele Yaka of Congo thus emphasizes the importance of appropriate sharing through the guiding concept of *ekila*. As Lewis (2008/9, 13) states, "For Yaka, people should be successful in their activities because nature is abundant. If they are not, it is because they, or somebody else, has ruined their *ekila* by sharing inappropriately." Significantly, "*ekila* regulates Yaka environmental relations by defining what constitutes proper sharing" (13). *Ekila* is ruined by an action such as not sharing hunted meat, being excessively successful and thus engendering envy, inappropriately sharing sexuality, or sharing laughter in such a way that the forest will not rejoice. By regulating potency through appropriate sharing, dynamic abundance is maintained for all. As Lewis writes, such culturenature ontologies and associated value practices have established a relationship with "resources" that has meant that Yaka people have "experienced the forest as a place of abundance for the entirety of their cultural memory" (13). This, again, is in rather stark contrast with modern discourses of resource scarcity and the associated competitive urgency to capture "values" in both extractive industry and conservation activity.[14]

Working in a different context again, anthropologist Eduardo Viveiros de Castro (2004) speaks of the similar *multinatural* "perspectivism" of cosmologies associated with peoples of the Amazon, a concept that currently is much celebrated by Bruno Latour (2004, 2010a). Viveiros de Castro posits perspectivism as the understanding that all beings share culture, kinship, and reciprocal relationships, their perspectives differing due to being seated in different bodily affects (or "natures"). Key aspects of this proposition are as follows: of an original culture that is disaggregated into different embodied perspectives; of all animals and plants being conceived as subjects/persons sharing a spirited hypostasis cloaked in different embodied perspectives; and of all embodiments as sentient, alive, and able to act with intentionality. Ecological relations thus are social relations, with all persons able to share and exchange knowledge. Communication and even transformation between such different embodied perspectives is an intrinsic possibility. These perspectives are understood in contradistinction to the naturalism of modernity, which proposes a shared universal Nature from which human culture and Reason rise and become progressively separate (see the critique in Gray 2002). Indeed, science becomes scientific when the world is decluttered of intentionality (Viveiros de Castro interviewed in Melitopoulos and Lazzarato 2012a, 4), such that the life sciences, on which modern conservation policy depends, propose a radically emptied encounter with nonhuman life. The "Amerindian" conception instead is that, "having been people [in the mythological past] animals and other species continue to be people behind their everyday appearance," endowed with the soul or spirit that personifies them (Viveiros de Castro 2004, 467). As such, "nonhumans," including ancestors and spirits, are attributed with "the capacities of conscious intentionality and social agency" (467). They are understood as subjects with empathically knowable and communicable subject positions that complexify possibilities for social and moral action.

Cognate culturenature orientations have been confirmed for me through ethnographic fieldwork since 1992 with people associated with the names Damara / ≠Nū Khoen and dwelling in northwest Namibia (also see Biesele 1993; Lewis-Williams and Pearce 2004; Low 2008).[15] I have written on this in the journal *New Formations* (Sullivan 2010a), and I reproduce some of that material here. This is a context where a rain shaman dances into trance and in this state of consciousness is able to climb a rope of light into a different but no less real world inhabited by the spirited beings that shape and form the life force(s) of daily embodied existence. Here he negotiates with the rain goddess |Nanus, seducing her to permit him to retrieve life-giving rain, which is then brought back with apparently real and celebrated effect (||Khumub et al. 2007). In this context, people can shapeshift into lions and other animals and be witnessed doing so, iterating the "reality of becoming-animal, even though one does not in reality become animal" (Deleuze and Guattari [1980] 1987, 273). Giant snakes, sometimes with antelope horns on their heads and quartz or lights in their foreheads, are known to roam the landscape, filling it with intense generative potency (Hoff 1997; Schmidt 1998; Low and Sullivan forthcoming). Here the most all-knowing deity is an insect—the praying mantis—who capriciously shapeshifts into and shares kin relations with many other animals, thus iterating the dynamic ambiguity of life itself as a force to be moved with in ways that maintain, rather than still, this movement (personal field notes; Biesele 1993). And here illness is carried and caused by wind, smell, and energetic arrows, with healing accomplished by the manipulation and alignment of energetic forms called |gais so that they stand up straight in the body (Low 2008). Culturenature assemblages of potency thus enfold human and nonhuman domains into endlessly dynamic connectivities, establishing mysteriously mutable relationships between what Occidental ontologies know as different orders of being (Biesele 1993; Power and Watts 1997; Low 2008). All of these phenomena, spoken of in contemporary times, sit within and affirm an old and broad KhoeSān conceptual world that speaks suggestively through the layers of rock art imagery, which is enormously prolific in southern Africa.

My final example embraces a quite different cultural context and is detailed in a 1986 article by Esther Cohen called "Law, Folklore and Animal

Lore," from which I will quote extensively.[16] Cohen describes the practice of "the criminal prosecution and execution of animals" in both secular and ecclesiastical courts of Western Europe in the later Middle Ages and the early modern period. She draws on legal anthropology and associated cross-cultural methodologies to assist with understanding the mutual social obligations that normatively bind animals and humans in these trials. Animal trials are first mentioned during the thirteenth century in northern and eastern France, from where they spread to the Low Countries, Germany, and Italy. They are documented in court records from the thirteenth to the eighteenth century, "reaching their peak of frequency and geographical scope during the fifteenth, sixteenth and seventeenth centuries" (Cohen 1986, 17). In them, sentences "were passed and executed in properly constituted courts of law by fully qualified magistrates, according to generally accepted laws," at the same time as being "an integral part of customary law" and owing "their continued existence partially to popular traditions and influences" (10). They generally followed two distinct procedures, secular and ecclesiastical. Secular procedures, for example, were "used to penalize domestic beasts that had mortally injured a human being," while ecclesiastical procedures were "employed to rid the population of natural pests that could not individually be punished" (10). Frequently, sentences were passed only after "ponderous debates and trials years long" (16). Here I provide some detail from a description of one of these proceedings, a trial of domestic animals in a secular court. My intention is to illustrate the seriousness with which nonhuman animals in these relatively recent European cases were attributed with subjectivity, intentionality, and personhood, leading to the animals' treatment as legal persons in the processes surrounding their trial and sentencing. These trials "differed as little as possible from human trials," usually involving appointment of an advocate for the defense of the accused nonhuman animal(s) (13).

Drawing on references from archival research, Cohen (1986, 10–11) writes:

In December 1457 the sow of Jehan Bailly of Savigny and her six piglets were caught in the act of killing the five-year-old Jehan Martin. All seven pigs were imprisoned for murder and brought to trial a month later before the seigneurial justice of Savigny. Besides the judge, the protocol recorded the presence at the trial of one lawyer (function unspecified), two prosecutors (one of them a lawyer and a councillor of the duke of Burgundy), eight witnesses by name, "and several other witnesses summoned and requested for this cause." Though the owner was formally the defendant, it is clear from the proceedings that he stood accused only of negligence and was in no danger of any personal punishment. Moreover he was allowed to argue in court "concerning the punishment and just execution that should be inflicted upon the said sow," if he could give any reason why the sow should be spared. The owner having waived this right, the prosecutor requested a death sentence. The judge, having heard all the relevant testimony and consulted with wise men knowledgeable in local law, ruled, according to the custom of Burgundy, that the sow should be forfeit to the justice of Savigny for the purpose of hanging by her hind legs on a suitable tree. The piglets created a more difficult problem as there was no proof that they had actually bitten the child, though they were found bloodstained. They were therefore remanded to the custody of their owner, who was required to vouch for their future behaviour and produce them for trial, should new evidence come to light. When the latter refused to give such a guarantee, the piglets were declared forfeit to the local lord's justice, though they suffered no further punishment. The court brought from Chalon-sur-Saône a professional hangman who carried out the execution according to the judge's specific instructions.

Cohen (1986, 11) explains that "the case of the sow of Savigny is typical in many respects of most secular animal trials. In the first place, it was held in Burgundy, one of the earliest areas to record such cases." In addition, "the defendant's porcine nature also recurred in a great many trials. Pigs, who seem to have accounted for the deaths of many unattended infants, were the most common culprits, but such trials also occurred throughout this time for homicidal pigs, oxen, cows, horses and dogs" (11). What is particularly relevant here is that "the trial is typical in its painstaking insistence upon the observance of legal custom and proper judicial procedure. This was neither a vindictive lynching nor the extermination of a dangerous beast. Other records mention, in addition to pre-trial imprisonment, the granting of remissions to wrongly accused beasts, the burning in effigy of a 'contumacious' [i.e., willfully disobedient] animal, and the public display of an executed cow's head" (11). Further, "where the hangman's bills are extant, they closely resemble those presented for the execution of humans" (12).

In her analysis, Cohen (1986, 15) notes that "the very existence of animal trials in Europe poses severe problems for the historian of Western culture" *because* "the practice runs counter to all commonly accepted conceptions of justice, humanity and the animal kingdom; and yet it survived and flourished for centuries." She writes that it is apparent that there are no clear distinctions between these domains in "the minds of medieval legists" (19). This is an ontological disposition that overlapped significantly with an emerging and elite modern rationalism regarding "the immutable categories of nature" and associated universal hierarchies (23–24), as well as with the radically different Cartesian notion that "animals are automata possessing neither sense nor feelings" (16).[17] For the medieval and early modern mind, the difference between "man and beast" instead "was functional, not causal: pigs or locusts who harmed man must alike stand trial in the interest of universal justice" (19). In parallel with the ethnographies of non-Western cultures discussed above, European animal trials thus seem to have "expressed a perception of law that held sway over the entire universe" for people who "viewed justice as a universal attribute, applicable to all nature" and in which "animals were neither insensate nor lacking in intent" (35–36).

In summary, these examples gesture toward an *amodern* "ontoepistemology" generating experiences of and dealings with "nonhuman natures" that depart radically from those empowered in the modern era.[18] They do more than simply suggest that nonhuman natures and objects are animate(d) actants producing effects and affects. Key additional themes emerge to stabilize the grid of this amodern episteme. "Nonhuman" entities are understood to embody variously different perspectives in a shared moral community of "persons," all of whom possess and enact intentionality that is communicable and knowable. The "social character" of relations between humans and nonhumans tends toward multiway economies of gifts, exchanges, sharings, and transformations between all persons (see Sullivan 2009; Haber 2012) and to mitigate against a commodity economy based on the creation and production of disembedded, pacified things (Viveiros de Castro 2004, 481–82). In addition, an array of "counterexistential" but actually commonplace experiences and ontological configurations permits transversal movements into other experiential domains populated by beings known and related with through millennia of dynamic biocultural concerns and desires.

Deleuze and Guattari ([1980] 1987, 309) write that what is valued in this "amodernity" is the ability and skill to improvise well with "what already is musical in nature."[19] Ontologically, this seems quite distinct from a modern imaginary that fixes nature and nature knowledge through surveys, measurements, maps, numerical models, and metrics (as discussed in Robertson 2012; Hannis and Sullivan 2012) and whose expert readers and constructors can be ordained to know their silenced constituents in advance (Castree 2006, 161). Improvising-with instead confers what Guattari ([1989] 2000, 21) refers to as the "significance of human interventions" in a context of an always and potently communicative nonhuman world that also is sentient and mind-full and asserts responsive agency. As Deleuze and Guattari ([1980] 1987, 258) state, "the plane of composition, the plane of Nature, is precisely for participations of this kind."

Becoming-Animist?

> *People call the soil mineral matter, but some one hundred million bacteria, yeasts, molds, diatoms, and other microbes live in just one gram of ordinary topsoil. Far from being dead and inanimate, the soil is teeming with life. These microorganisms do not exist without reason. Each lives for a purpose, struggling, cooperating, and carrying on the cycles of nature.*
> —MASANOBU FUKUOKA, quoted in S. H. Buhner,
> The Lost Language of Plants

> *The disenchantment of the world is the extirpation of animism.*
> —T. ADORNO and M. HORKHEIMER,
> Dialectic of Enlightenment

> *Today, it seems interesting to me to go back to what I would call an animist conception of subjectivity.*
> —FÉLIX GUATTARI, quoted in A. Melitopoulos
> and M. Lazzarato, "Assemblages: Félix
> Guattari and Machinic Animism"

The understandings suggested in the previous section propose cogent "counterlogics" and praxes regarding nature/culture relationships that open the black box of mute nature proposed by modernity's great divide (see Latour 2004), the circulating abstractions of which infuse the current conceptual and policy paradox of "green growth." Such counter-culturenature ontologies may indeed be among the social forces that can be mobilized and affirmed today in (re)configuring, (re)composing, (re)embodying culturenature relationships that are enlivened in support of the flourishing of life's diversity (Sullivan 2010b), thus curtailing the modernist project of severed relationships (Hornborg 2006). Animist moral economies propose conceptual and ecoethical space for the dynamic sustenance of relationships between diverse entities, with all acting to play a part in this maintaining "sustainability" (Descola and Pálsson 1996a, 14; Harvey 2005; Bird-David and Naveh 2008; Schwartzman 2010, 322). It is this "power effect" that makes animist culturenature ontologies worthy of engagement, given the Anthropocenic juncture at which collectively we find ourselves.

But it can be difficult to speak of such animist counterlogics and ontologies within academia and other modern institutional contexts. This is due both to the necessary systemic "epistemological purification" of such amodern knowledges for the consolidation of modern categories (Descola and Pálsson 1996a, 8) and to the mirror of falling prey to "'archaic illusion,' where moderns . . . nourish their fantasies about the primitive other, mysterious communications, mimetic contagions, spirits, enchanted nature, and so forth" (Franke 2012b, 21, after Taussig 1987). Anthropologists are specifically hampered by a charge that in speaking of animist culturenature counterlogics we might iterate a romantic and nostalgic construction of indigenous peoples as living in some sort of unreachable and ahistorical harmony with a spirited nature. Kuper (1993), for example, argues that such a romanticism, and a delineating of "indigenous peoples" and affective relationships with "the environment" more generally, effects a problematic "return of the native" in anthropology. He suggests that this echoes earlier colonial characterizations that served to denote and demote the "other" and that made possible the displacements and violences enabling the reconstitution of people and nature as labor and property.

There is a danger here, however, of throwing the baby out with the bathwater (Sullivan 2006a), by implying that it is only legitimate to understand relationships between culture and nature from the perspective of the ontological bifurcation between them—via which "nature" can be peered at from the culture side of the fence, and any refraction of this divide, in terms of where subjectivity, agency, and intentionality might be located, becomes subject to dismissal. It also imputes a valorization of essentialized identities, as opposed to a curiosity regarding different culturenature praxes and their productive effects. In other words, it is not that the animist culturenature conceptions, experiences, and value practices explored above are interesting because they might be those learned from indigenous peoples (the European example from Cohen [1986] in any case destabilizes this pattern here). It is because the conceptions and praxes themselves might have effects that are relevant for coming to terms with being human in the Anthropocene, as well as for making choices regarding subjectivity that might be better calibrated with life's diversity.[20]

Indeed, the current global socioecological cul-de-sac in which collectively we find ourselves suggests that continued dismissal of such different culturenature ontologies is a luxury we can ill afford. As ecologist Richard Norgaard (2010) describes, in shoe-horning our understandings of nature such that the only valid terms and concepts are contemplated, objectified, and monetary ones (whether metaphorical or as newly devised and tradable commodities), a foundational contraction of possibilities is occurring. Options for different socioecological praxes are being foreclosed, even as a new frontier for capital investment in nature conservation is composing new "socionature" and "world-ecology" possibilities.[21]

This, then, is a proposal for a positive and refracting dialectics (Ruddick 2008; Latour 2010a; Gibson-Graham 2011) that is inspired by animist ontoepistemologies so as "to undo the very 'alienation' that capitalist modernity induces" (Franke 2012b, 21). It is a proposal for engagements that mobilize knowledge of the cultural and historical particularities that have silenced "nonhuman nature" and diverse biocultural knowledges so as to resuscitate and affirm immanent "counterlogics" and praxes that might bring socionature arrangements "back to life." Bennett (2010, 14) affirms that "the starting point of ethics is . . . the recognition of human participation in a shared, vital materiality" (see also Goldstein 2012). The culturenature ontologies of other(ed) cultural perspectives offer much for the guiding of such recognition. At the same time, their (re)countenancing requires both considerable decolonization of the orders of knowledge sustaining modernity and a turning to face the systemic violences with which these orders have been established and maintained. Nonetheless, and to invoke a hopeful Foucault ([1976] 1998), since the strategic relationships, practices, and discourses that become empowered also always contain their own "gaps"—their own possibilities for breakdown, subversion, and reconstitution—a corresponding potential exists for interventions that exploit these contradictions and ambivalences.

In moving from critique toward insertions that may refract and reconstitute, however, "we" also need to have something different to say. In the spirit of "ambitious naiveté" (Bennett 2010, 19), I hope here to have brought in some suggestions for ways in which culturenature relationships might be thought and practiced differently and thereby to have provided elements of something different to say and do.

Postscript: Ethical Gestures toward a Transcendental Immanence

> The transcendental field is defined by a plane of immanence, and the plane of immanence by a life. . . . Transcendence is always a product of immanence.
>
> —G. DELEUZE, *Pure Immanence*

The world is holy! The soul is holy! The skin is holy!
The nose is holy! The tongue and cock and hand
and asshole holy!
Everything is holy! everybody's holy! everywhere is
holy! . . .
Holy the supernatural extra brilliant intelligent
kindness of the soul!
— ALLEN GINSBERG, "Footnote to Howl"

Nature never became a toy to a wise spirit.
— RALPH WALDO EMERSON, *Nature*

Even the difference between transcendence and immanence
seemed to be beyond them.
— BRUNO LATOUR, *On the Modern*
Cult of the Factish Gods

On closing this essay, I realized I had made an omission. I had overlooked making any gesture toward considering and locating "the sacred" and its significance in the animist culturenature ontologies engaged with above. This is curious, since my sense is that animist tendencies, with a healthy force of humor (Willerslev 2012), centralize the sacred in conceptions and constitutions of culturenature, with potent ecoethical effects. The sacred is everywhere present: as the soul connecting relational entities of different form (Buhner 2002; Harvey 2005); as a sanctioning of the gaps in knowing generated by the experience of mystery that thereby emerges (Wheeler 2010, 44); and as the lived relationships via which each being in "the cosmic community of beings . . . is bred, grows, reproduces and dies" (Haber 2012, 5).[22]

Reflecting on why this omission occurred brings to mind a series of constraining and silencing trajectories. Of a millennia-old capture of the sacred by priestly castes tasked with mediating between a sanctified heavenly realm inhabited by a distant, individualized, and judging God and a populace of lesser mortals denied legitimate experience of "the divine" (Young 2011). Of the similarly transcendent expert knowledge and religious fervors (Wheeler 2010, 37) of priestly castes of scientists, entrepreneurs, and politicians whose choices are elevated in Man's continuing dominion over Nature. And of a simultaneous historical and contemporary denial of sacred presence in the ordinary natures of everyday life, combined with the occlusion of commonsense knowledges and practices of those experiencing as well as instrumentalizing this immanent presence (Federici 2004). Through this nexus of circumstances the sacred becomes set apart from the earthly and fleshly germinative plane of immanence, such that participation in, and engagement with, earth and body are devalued. The sacred as transcendent experience has tended to be seen in contradistinction to the immanent sphere. A transcendent God is both beyond the limitations of the material universe and beyond knowing by nonspecialist humans, not to mention being intrinsically unavailable to creatures deemed made less closely in the image of Him.

But of course, this is not the only way in which the sacred might be conceived, as expressed by poets, mystics, shamans, and critics of all times and cultures. In his 1836 essay *Nature*, the North American poet and essayist Ralph Waldo Emerson destabilizes this sense of a transcendent sacred realm that is unknowable or unreachable without the mediation of empowered experts through an exposition that paradoxically became known as "Transcendentalism." In this, "nature" is deemed poetically knowable by the most innocent of minds through the attunement of the senses between inner and outer worlds. He speaks of "an occult relation between man [*sic*] and the vegetable" in which "I am not alone and unacknowledged. They nod to me, and I to them" (Emerson [1836] 1985, 6). Emerson's "Transcendentalism" affirms a pantheistic sacred immanence infused with an "ethical character" that "so penetrates the bone and marrow of nature, as to seem the end for which it was made" (28). The sacred, as transcendent and intuitive experience (Wheeler 2010, 37) and as entranced state of consciousness (cf. Sullivan 2006b; Fletcher 2007, and references therein), thus is immanent in a nature, the generation of which humans and other persons are part (Ingold 2006). Transcendent sacred experience is an ordinary possibility for the human by virtue of being a facet of nature's immanence that can also know and open to the other aspects of nature's diverse embodiment (Bateson and Bateson 2004, discussed in Wheeler 2010). As Hepburn (1984, 184, quoted in Curry 2008, 64, emphasis added) states and as echoed later in the quote by Deleuze that opens this section, "there is no wholly-other paradise from which we are excluded: *the only transcendence that can be real to us is an 'immanent' one.*"

A sense of this commonality perhaps is present in the ethnographic examples above. Viveiros de Castro (2004, 464) describes an ontological "state of being where self and other interpenetrate, submerged in the same immanent, presubjective, preobjective milieu." The hypostasis of embodied difference thus is an all-pervasive, connecting, and communicative vitality—an ontological primacy of animacy, as Ingold (2006, 10) puts it. Arguably, the instrumentalizations of life and landscapes associated with monotheistic doctrine, Enlightenment thought, and the rise of modern capitalism are effected in conjunction with the enforced denial and systemic disruption of this embodied immanence (Weber [1904] 2001; Merchant [1980] 1989; Federici 2004). This is a constraining gendered dynamic too, in that the subject position of Western transcendental/Enlightenment philosophy—the "Father-*logos*" that "claims to be the overall engenderer compared to mother-nature," sets up "a transcendence corresponding to a monosexual code" (Irigaray 1997, 314). This "Law-making-God-the-Father" equates "to an absolute transcendence only insofar as it is appropriated to male identity," in the meantime ensuring that "everything that is of the feminine gender is . . . less valued in this logic because it lacks any possible dimension of transcendence" (314). When this includes a feminized earth, the feminized values of the body and of the (indigenous) natural become discarded and violated (as documented in brutal detail in Merchant [1980] 1989; Taussig 1987; and Federici 2004). This generates "the ecofeminist insight that there is a relationship between the subordination of women and the exploitation of nature" that is extended to indigenes, configured conceptually as similarly close to nature (Mellor 2000, 107). It is associated with a patriarchal circumstance in which "dominant men" are "above nature (transcendent)," while "women are seen as steeped in the natural world of the body (immanent)" (111; see also Sullivan 2011c). Mary Mellor (2000, 117) thus urges a conceptualization of "human envelopment in 'nature' as a material relation, an immanent materialism, that is the historical unfolding of the material reality of human embodiment and embeddedness within its ecological and biological context."

But perhaps it is the experience of this material immanence as also a transcendent experience of the animate embodied sacred that enhances ecoethical behaviors? This, then, is an affirmation of the ethical praxes that might be engendered by the notion of a "transcendental immanence," arising both from the "transcendent experience" of the inviolable sacred as immanent or in-dwelling in all entities and relationships; *and* from the a priori possibility that such experience is part of the immanent "toolkit" of the embodied "human condition" (see Spinoza [1677] 1996). It is based on the proposition that when sensual and communicative vitality is known as shared by and pervading all entities, it arguably (and hopefully) becomes harder to make choices that violate socioecological integrity. As Bennett (2010, 14) iterates, "the ethical task at hand here is to cultivate the ability to discern nonhuman vitality, to become perceptually open to it." In this vein, then, and in solidarity with a growing number of authors (see Merchant [1980] 1989; Abram 1996; Roszak [1992] 2001; Buhner 2002; Harvey 2005; Ingold 2006; Curry 2011), a revitalized experience of living in embodied relationship with a communicative and animate nature is a necessity if current alienations and violences are to transmute into democratic and vivacious socioecological sustainabilities. With Allen Ginsberg in the provocative quote above, it is a reminder that everywhere, everything, and *every body* is holy and can be reimagined, experienced, and treated ethically as such.

Notes

When this chapter was first presented at the conference "Nature™ Inc.: Questioning the Market Panacea in Environmental Policy and Conservation," it was accompanied by a short film, which can now be viewed online (http://siansullivan.net/talks-events/). This chapter is dedicated to the memory of Kadisen ‖Khumub, rain-shaman of the Etosha Hai‖om and road laborer for Etosha National Park, Namibia. In *Vibrant Matter* Jane Bennett (2010, 21) writes that "an actant never really acts alone"—that "agency always depends on the collaboration, cooperation, or interactive interference of many bodies and forces." In the years that this piece has been gestating, I have been sustained by many such inspirations and frictions. My appreciation goes in particular to my partner, environmental philosopher Mike Hannis, and to friends, colleagues, and human and "other-than-human" collaborators in the Viva, Nature™ Inc., Movement Medicine, and Kings Hill communities. All errors remain mine alone.

1. This is my adaptation of the Sumerian creation myth of around 2000 BCE, later retold as the Babylonian story *Enuma Elish*. Summarized in Willis and Curry (2004) and Young (2011).

2. I use the terms "other-than-human nature(s)" and occasionally "nonhuman nature" and "more-than-human nature" when referring to organisms, entities, and contexts other than the modern commonsense understanding of the biological species *Homo sapiens*. As highlighted in this chapter, however, these terms are already culturally embedded and constructed. For cultural contexts where the "nonhuman" is "personified" and there is a tendency toward the assumption of one humanity and many different embodied perspectives, these terms are problematic and even nonsensical. In the ontological domain of shamanic "perspectivism," for example, there are no "nonhumans" (Viveiros de Castro 2004).

3. As framed, for example, by the EU- and UN-supported TEEB (The Economics of Ecosystems and Biodiversity) program, on which see Sukhdev (2010), and by the United Nations Environment Programme (UNEP 2011). For critical engagement, see Büscher et al. (2012).

4. Note that this is a move that echoes the rise of the signifier of equilibrium in colonial ecology and the imperial tendency to view ecosystems of the "periphery" in terms of a definable and desirable climatic climax, with anything different to this classed as degradation through irrational (indigenous) use practices (see, e.g., Anker 2001; Sullivan and Rohde 2002).

5. Commodity fetishism emerges in Marx's writings to clarify "the relationship between exchange value and use value as it is embodied in the commodity" (Holert

2012, 4), whereby the value of an object is seen as residing in the thing itself in a manner that obscures and thus alienates the labor (and nonhuman life) from which it is made (Graeber 2011, 65). The systemic screening-out of materiality and labor relations from commodity production and consumption under capitalist commercialization creates a logic that endows commodities with something akin to a soul, wherein they appear to assume human powers and properties and thus act to satisfy wants. Marx (1975, 189, quoted in Nancy 2001, 4) derived his theory of commodity fetishism from interpretations of the fetishistic abstractions of objects amongst noncapitalist societies at the colonial frontier, stating that "fantasy arising from desire deceives the fetish-worshipper into believing that an 'inanimate object' will give up its natural character in order to comply with his desires." He extended this to the abstracted commodities and currencies produced under capitalist relations of production, including money—hence "the magic of money" (quoted in Nancy 2001, 5). A corresponding attribution of agency to capital, capitalism, and markets has led Michael Taussig (1987) to speak of a "capitalist animism" (see discussion in Holert 2012; also Jones 2013). A "postcapitalist animism" (Holert 2012) instead might note that a modern removal of subjectivity and intentionality from nonhuman entities was itself an historically embedded discursive move that facilitated the creation of a scientifically knowable, exploitable, and tradable world of objects. Marx's ([1853] 1962) understanding of "primitive" fetishistic practices and "the brutalising worship of nature" derives from this context. While foregrounding the "truths" that are screened out by the activities of commodities and capitalisms, it is worth noting, then, that the concept of "commodity fetishism" is steeped in particular understandings of the "fetish" as a component of "primitive" and animist thought and is associated with a broader modern dismissal of amodern animist ontologies as "savage" and irrational. This chapter seeks in part to reclaim amodern animist ontologies from such dismissals, noting that in any case the apparently exterior "matters of fact" and commodity objects of the modern are themselves fetishized "factishes," as Latour (2010b) puts it—brought into being through human work but charged with acting from a distance as exteriorized facts animated technically and socially with authoritative, objective power. We may never have been modern, because we are all fetishists, endowing the materialities we create and with which we are embedded the powers to shape our actions, choices, and affects.

6. See the Bank of Natural Capital website established by TEEB at http://bankof naturalcapital.com/.

7. REDD+ refers to the United Nations program for Reducing Emissions from Deforestation and Degradation in Developing Countries (http://www.un-redd.org).

8. I have observed the immense work to raise sources of finance for biodiversity conservation by policymakers associated with the UN Convention on Biodiversity (CBD) through being an invited participant in meetings held by the UN Secretariat of CBD and partners on "Scaling Up Biodiversity Finance" (in Quito, Ecuador, March 2012, http://www.cbd.int/financial/quitoseminar/) and "Ecology and Economy for a Sustainable Society" (in Trondheim, Norway, May 2013, http://www.dirnat.no/tk13/).

9. As, for example, in the binding of distant localities to financialized trade in carbon and associated option and futures exchanges, as well as in weather derivatives and various environmental futures and derivatives (see the emissions trading page of the Intercontinental Exchange [ICE], https://www.theice.com/emissions.jhtml, and discussion in Böhm and Dabhi 2009; Cooper 2010; Randalls 2010; and Lohmann 2012).

10. See, for example, WildEarth™ at http://www.wildearth.tv/, accessed October 23, 2011.

11. For examples of such environmental conservation markets, see Carroll, Fox, and Bayon (2008) and Briggs, Hill, and Gillespie (2009); for discussion, see Robertson (2004, 2006); Morris (2006); Robertson and Hayden (2008); Pawliczek and Sullivan (2011).

12. In invoking "place" and "emplacement" here, I follow Ingold's (2005, 507) conception that "places are not static nodes but are constituted in movement," in comings and goings and through embodied actions and perceptions, all of which necessitate movement in conjunction with an always moving milieu of nonhuman presences (see also Abram 1996, 65). This is not to say, of course, that times of disagreement, bloodshed, and warfare do not occur in these circumstances (see Rival 1996).

13. On bringing diverse biocultural and spiritual values into modern conservation praxis, see the special issue of the *Journal for the Study of Religion, Nature and Culture* titled "Indigenous Nature Reverence and Nature Conservation," introduced and edited by Snodgrass and Tiedje (2008), and the Biocultural Community Protocol Toolkit at http://www.community-protocols.org/toolkit, accessed April 24, 2012.

14. This competitive urgency to capture "values" in both extractive industry and conservation activity is also increasingly compounded by a global movement in which the offsetting of impacts from the former may enhance the scarcity and financializable "value" of the latter (e.g., see Seagle 2012; Sullivan 2013a).

15. The symbols used here reflect the standard orthography for KhoeSān languages used to denote click consonants.

16. Thank you to Martin Pedersen for drawing my attention to this paper.

17. In *Discourse 5* of René Descartes's *Discourse on Method* ([1637] 1968, 75–76), he writes of animals that "they do not have a mind, and . . . it is nature which acts in them according to the disposition of their organs, as one sees that a clock, which is made up of only wheels and springs, can count the hours and measure time more exactly than we can with all our art." Other authors argue against the thesis that Descartes considered animals to be incapable of feeling while affirming his insistence on animals as automata, possessing neither thought nor self-consciousness (Harrison 1992, 219–20). It is telling that the emerging Cartesian vivisectionists "felt compelled to sever the vocal chords of the dogs whose living anatomy they explored," thus performing "their modernist task" only after having literally silenced their subjects in the endeavor of transforming them into objects of study (Hornborg 2006, 24).

18. By *amodern* "ontoepistemology" I mean reasoned knowledge flowing from particular cultural and historically situated assumptions regarding the nature of reality and the methods through which, given these assumptions, it is possible to know this. I derive the term "ontoepistemology" from Jones (1999). On the connected understanding of episteme as the cultural and historical fabric that shapes and determines what it is possible to know, see Foucault ([1966] 1970). Foucault uses the term "episteme" to describe the assumed or a priori knowledge of reality—the knowledge that is taken as given—that infuses and permits sense making to occur in all discursive interactions flowing from and reinforcing a historical period or epoch. This is similar to an understanding of "culture" as the shared norms and values that infuse and produce community in all spheres of praxis and language. An episteme thus guides and influences the social production of discourses—or empowered knowledge frames—that at the same time

iteratively reproduce what epistemologically is, and becomes, shared as self-evident about the nature of reality.

19. For more on improvisation as a dynamically sustaining praxis, see Gilbert (2004) and the edited volume by Ingold and Hallam (2007).

20. Compare Guattari's ([1989] 2000, 19–20) differentiated and multiplicitous "ecosophy" as "an ethico-political articulation" between the three ecological registers of "the environment, social relations and human subjectivity" that reimbeds relationships between interior (subjective) and exterior (social and environmental) potencies.

21. The term "socionature" is borrowed from Swyngedouw (1999), and "world-ecology" is from Moore (2010). It seems hard to find a term in English that unclumsily expresses connectivity between human and "other-than-human" worlds. It seems important to do so, however, so as to keep affirming connections and correspondences between these worlds. After all, no individuals of any species, including our own, are actually able to exist in a state of disentanglement from other species (see Ingold 2010).

22. The current government of Bolivia has integrated a conception of sacred within its legal framework for *buen vivir* (living well), as in chapter 2, article 4(2) of the "Framework Law of Mother Earth and Integral Development for Living Well," which states that "the environmental functions and natural processes of the components and systems of life of Mother Earth are not considered as commodities but as gifts of the sacred Mother Earth."

The Limits of Nature™ Inc. and the Search for Vital Alternatives

Wolfram Dressler, Bram Büscher, and Robert Fletcher

This book began with the suggestion that "nature trademarked incorporated" stands as an apt metaphor for conservation in our times. The premise argued that in the neoliberal age, new frontiers of environmental conservation have become inextricably bound, defined, invested, and reinvented in terms of the interrelated dynamics of commodification, competition, financialization, and market disciplining that were less dominant in earlier management regimes. As a concept and reality, Nature™ Inc. has set out to define and condition not only the design and substance of conservation planning and practice but also the fundamental values, beliefs, and assumptions that underlie a range of conservation frameworks and practices across different scales and, in the process, has "refurbished" the sociocultural dynamics, livelihoods, and landscapes that support the prevailing economic system (and the ecologies underpinning this system).

As each chapter reveals, the enactment of the concept has shown how social and cultural relations embedded in land and lifeways become hollowed out and refurbished within and through market mechanisms aiming "to transcend the conservation of particular in situ natural resources to allow for the abstraction and circulation of 'natural capital' through the global economy," as described in the introduction. The outcomes of this process have had profound discursive and material reach and impact, strongly influencing perceptions and representations of human relations with nonhuman natures. As noted in the introduction, with so much at stake, it remains critical to continue to investigate how neoliberal conservation actively reshapes human-nature relations in the context of several centuries of capitalist development.

While acknowledging the need to keep problematizing the neoliberal order in the face of its increased promotion in global environmental governance arenas, however, we agree with a growing chorus of thinkers, practitioners, and activists that it is necessary to also promote serious discussion of alternative forms of (re)production and ways of being that go beyond Nature™ Inc. in all its proliferating forms. This entails, first, moving the debate on the politics and political economy of conservation forward by outlining and encouraging new theoretical perspectives on the process; and second, reflecting on and informing empirical practice directed toward non- and/or postcapitalist spaces and possibilities (Gibson-Graham 2006; Gibson-Graham, Cameron, and Healy 2013). The two aims are of course bound together and cannot be treated adequately in a short concluding chapter. This is not our aim. Rather, we build on the book's content by outlining what we believe are key themes, issues, directions, and initiatives involved in starting to move discussions and practices beyond neoliberalism in pursuit of what we and the editors of the series this book is part of call "Vital Alternatives."

We begin by recapping some of the key issues raised in the preceding chapters, contending that what our critiques and analyses add up to, essentially, is the recognition that there are indeed clear limits to the capacity of a neoliberal approach to effectively commodify biodiversity conservation. This grounds our call for consideration of Vital Alternatives, where we outline a variety of emerging perspectives and possibilities that may help us to transcend Nature™ Inc. in pursuit of a world beyond the impoverished (and impoverishing) promises of neoliberal capitalism. We conclude by outlining a brief manifesto for the new book series of which this volume is the first installment. In this sense, both the present volume and the overarching series function as a broader call for new discussions and practices involving conservation and human-nature relations as we move deeper into the twenty-first century.

The Limits of Nature™ Inc.

As the chapters in this volume reveal, neoliberal conservation is highly problematic for a number of reasons that for the sake of brevity can be summarized under material and discursive dimensions. Materially, a neoliberal approach to environmental governance commonly contributes to economic and other inequalities due to its propensity to redistribute "upward" rather than "downward" (Dressler 2009; McAfee 2012a, 2012b). At the same time, neoliberalism's reliance on economic growth means that it necessitates environmentally destructive dynamics and activities, thus often forcing into opposition the very conservation and development concerns it ostensibly seeks to reconcile (Fletcher 2012c). In addition, provision of economic incentives to encourage conservation can backfire when those incentives are meager or unevenly distributed. Hence, West (2006, 185) highlights a situation in which disagreement concerning ownership of a particular tree housing a harpy eagle valued by foreign conservationists led to one disputant chopping down the tree, cautioning that a market approach "may well lead to environmental destruction instead of environmental conservation." At the same time, the failure to develop tourist markets for women's crafts was "making them hostile to the practice of 'conservation'" (207).

Moreover, an incentive-based approach has difficulty countering large-scale activities such as mining and logging since such activities are already highly profitable and their opportunity costs often preclusive, given the limited resources market-based conservation mechanisms are able to marshal (Fletcher 2012c). Relatedly, in other instances, resource extraction and market-based conservation (e.g., ecotourism) are increasingly drawn together as associated interventions, where the latter are supposed to (but in practice often do not) offset negative externalities and so reinforce the growth potential of the former (Büscher and Davidov 2013). Finally, it is questionable to what extent common faith in the potential of undiscovered commodities for which sustainable local markets can be developed (e.g., via bioprospecting; see Neimark 2012) is in fact realistic.

In terms of discursive dimensions, neoliberal conservation's reduction of a highly complex "nature" to single-dimensional "natural capital" evinces a striking "cultural poverty" (Sullivan 2009), foreclosing the far different means of valuing "natural resources" found beyond the realm of monetary exchange value (Singh 2013; Sullivan, this volume). Likewise, neoliberal conservation's characteristic promotion of successful "win-win-win" outcomes in the interest of satisfying donor and business partner expectations frequently conceals the contradictions and discrepancies operating beneath this rhetoric (Büscher et al. 2012; Fletcher, this volume; Lohmann, this volume). Further, virtual "nature" encounters and creative consumption encouraged by celebrities and business-centered conservation (e.g., buying a Happy Meal to help endangered species) functions as a form of commodity fetishism obscuring the often-dubious results of such strategies in practice (West 2006, 2010; Igoe 2010, 2013a; Igoe, Neves, and Brockington 2010; Brockington, this volume).

Obviously, these two dimensions of neoliberal conservation's problematic nature are only schematically presented and cannot be treated adequately in several short paragraphs. To do them justice would entail repeating the theoretical connections, ethnographic relations, and analytical linkages (re)established and presented by the contributors of this volume in response to the common neoliberal practice to "actively obscure" the social, cultural, and political connections that give societal and ecological diversity its substance, depth, and vibrancy — the basis from which alternatives may emerge as solutions (Duggan 2012). Indeed, the chapters were written against and in the face of neoliberal "master terms and categories" that "do not simply describe the 'real' world but rather provide only one way of understanding and organizing collective life . . . hiding stark inequalities of wealth and power and of class, race, gender and sexuality" (5).

As the consolidation of neoliberal conservation, Nature™ Inc. yields the same logic to organize and structure ideals, beliefs, and actions of transnational capitalist elites, practitioners, and resource users to retain more with less, but usually at the expense of the majority who are excluded from these relations (Sklair 2001). The main message here is that the homogenizing tendencies of Nature™ Inc. emerge from the simultaneous disciplining of beliefs and ideals and the intensification of actions and output, effectively narrowing fields of vision and production in line with restrictive means and ends. This stems from the broader governance structures and technologies of control — payments for ecosystem services, carbon monitoring, etc. — that intersect and strengthen preexisting capitalist relations of production, exchange, and consolidation of resource commodities, whether tangible or intangible (e.g., carbon trading intersecting with "boom crop" production). Indeed, the diverse theoretical and empirical case evidence that the chapters of this book bring to bear on the above exemplifies the ubiquitous discursive and material impact of neoliberalism in even the remotest of frontiers — that is, the ways in which both neoliberal economics and governance work within and through society to discipline and streamline thought and action among citizens (see, among others, the chapters by Matose, Wilshusen, Dressler, MacDonald and Corson, and Sullivan).

The extent and scope of such issues suggests that there are clear and immanent limits to the efficacy of Nature™ Inc. to achieve conservation on a significant scale. This is a conclusion to which several recent analyses are beginning to point (Arsel and Büscher 2012; Wilshusen, this volume; Fletcher, this volume; McAfee 2012a, 2012b). It is in line with Peck's

(2010, xiii) overarching assessment that neoliberal policies in general tend to evince limited efficacy in practice, functioning instead as "repeated, prosaic, and often botched efforts to *fix* markets, to build quasi-markets, and to repair market failure."

If this assessment is correct—and we believe it is—then it becomes all the more urgent to challenge the continued hegemony of Nature™ Inc. and open space for the vision and practice of more equitable and efficacious approaches to managing human-nonhuman relations. This is central to the search for Vital Alternatives, intended to be understood in two inter-related senses: (1) to designate the urgent necessity of developing viable alternatives to the current neoliberal order, particularly in the face of the growing environmental/economic crisis (Büscher and Arsel 2012; Fletcher and others, this volume); and (2) to emphasize that these alternatives must be founded on a more expansive understanding of the value and politics of life, both human and nonhuman, than capitalist economic rationality affords (Gibson-Graham 2006; Igoe, this volume; Sullivan, this volume).

Insofar as the contributors to this volume critically engage neoliberal conservation and other market-based governance processes, some offer tangible examples of practices falling to some degree outside these bounds. Yet few of us provide truly new or Vital Alternatives to Nature™ Inc.'s tendency to control, hollow out, and impoverish socioecological systems. As such, we further our discussion, analysis, and conclusion toward tangible steps in this direction, paralleling Hall, Massey, and Rustin's (2013, 18) Kilburn Manifesto *After Neoliberalism*, in which they argue that "the neoliberal order itself needs to be called into question, and radical alternatives to its foundational assumptions put forward for discussion. Our analysis suggests that this is a moment for changing the terms of debate, reformulating positions, taking the longer view, making a leap."

What, then, are some of the parameters for this leap with respect to neoliberal conservation? We outline below a variety of emerging approaches to addressing this question, along with others that we feel hold potential for future elaboration.

Vital Alternatives

Discussions about alternatives to capitalism are, of course, long-standing, including those that focus specifically on the environment. As we cannot do justice to these discussions here, we aim to simply give some examples of work we believe may specifically help to go "beyond neoliberal conservation." To explore "vital alternatives," after all, is to keep discussions, visions, and vistas of other, noncapitalist ways of being, seeing, and conserving alive and "vivacious" while at the same time showing their crucial importance. In doing so it matters less—for the moment—whether we fully agree with each of the alternatives we discuss, though it obviously remains vital to also continue scrutinizing and critically debating any proposed alternative. By simply giving an overview and acknowledging the importance of some discussions, we want to show that these are indeed vital, as in "lively," and that they deserve to be built upon—critically *and* constructively—in future work in this series and beyond.

A good starting point for discussions of "postcapitalism" is the work by Gibson-Graham (2006), including recent writing by Gibson-Graham, Cameron, and Healy (2013), in which they argue that we must "take back the economy." The authors suggest that this involves sustained "revisionist thinking" and action that can create alternative spaces within which to first buffer and then reverse the destructive nature of neoliberal practices. While these and other authors acknowledge that the neoliberal economy is one made up of decisions and choices that become normative common sense, they note that, despite this, it continues to be based on fallible human discourse and action that can be challenged and transformed. In other words, there is nothing inevitable about the long-term presence and impact of neoliberalism (see also Hall, Massey, and Rustin 2013). New and important perspectives are thus emerging not only to critically engage neoliberalism in advocacy writing but also to inform action in terms of ethical practice, the equitable distribution of surplus, and respectful and conscientious relations and consumption, as well as invest in protecting and/or revitalizing the commons (Gibson-Graham 2006; Gibson-Graham, Cameron, and Healy 2013; Wolff 2012).

Other work takes a more geographical turn in tackling capitalism's structural features and ingrained modes of power and being. David Harvey's *Spaces of Hope* (2000) is a good example, while Neil Smith's (2008) work has led to a flurry of analyses trying to understand how we could "produce" nature and space differently. Yet others explore how we can subvert and change "high-technology capitalism" to make space for "communications commons" focused on the "public financing of a multiplicity of decentralized but collectively or cooperatively operated media outlets, licensed on the basis of commitment to encouraging participatory involvement in all levels of their activity" (Dyer-Witheford 1999, 204, and see esp. chap. 7).

When these (and other) alternatives to capitalism are extended to alternatives to neoliberal conservation, one can take further inspiration from the many debates that explicitly link capitalism and ecology. These, too, are very diverse and range from radical Marxist socialism (Williams 2010; Magdoff and Bellamy Foster 2011), strategies for economic "degrowth" (Kallis 2011), anarchist geographies (Springer et al. 2012), and less radical "steady state economics" (Dietz and O'Neill 2013; Czech 2013) to "bioregional" economies (Scott Cato 2012). Moreover, there are calls for "living with" biodiversity (Turnhout et al. 2013) and emphases on affective hope and ways of relating with nonhumans that are different from the destructive capitalist ratio (Sullivan, this volume; Singh 2013). Allied with this are growing calls for legitimation of nondualist ontologies that, for example, adopt the perspective of "dwelling" (Ingold 2000) or embodiment (Valera 1992), or the increased attention devoted to Amazonian cosmovisions, among many others (Descola 2005; Viveiros de Castro 1998, 2004). These approaches work best in particular contexts, reflecting the unique natures and diversity in which they are immersed, but when appropriately linked, in concept and action, they offer tangible and inspiring alternatives to Nature™ Inc. We do not consider these clear or straightforward alternatives, for blueprints for action do not exist and indeed go against the needed diversity of democratic and socially just conservation spaces that we envision.

Toward a Research and Practice Agenda around Vital Alternatives

Overall, critical inquiry in the social sciences—spanning disciplines from anthropology to pedagogy—has perhaps a greater responsibility than ever before to move critique into vision and action by allying itself with movements, collectives, and interventions both "on the ground" and "in the mind"—the means by which diversity might (re)sustain and reinvigorate itself (against homogenizing tendencies) and so create spaces for Vital Alternatives. Movements throughout the "global South," particularly in South and Central America, have fused critical intellectual thought and writing with creative resistance in order to forge relatively autonomous spaces for thinking and acting out against the "burden of neoliberalism" (see, e.g., Arsel 2012; Arsel and Angel 2012; Grugel and Riggirozzi 2012). As Osterweil (2004, 8) describes, "asserting and creating . . . other ways of being in the world, these movements rob capital [or the state] of its monopoly and singular definition of time, space and value" and simultaneously help to "furnish new tools to address the complex set of problematic power relations" that confront and define the everyday realities and thus social and ecological existence of both conservationists and the rural poor. At their best, these movements reveal and produce "landscape[s] of radical heterogeneity populated by an array of capitalist and non-capitalist enterprises; market, non-market, and 'altermarket' transactions; paid, unpaid, and alternatively compensated labor; and various forms of finance and property—a diverse economy in place" (Gibson-Graham 2011, 2).

The new book series of which this volume is the initial installment seeks to contribute to this broader project, taking as its central theme critical green engagements, with specific attention to and articulation of Vital Alternatives to the growing green economy paradigm. The editors and authors contributing to this volume have sought to critically engage the theory, concept, and practice of neoliberal conservation not simply for the purpose of problematizing Nature™ Inc. but to illuminate spaces within which new debates and actions can proceed to help move conservation back to investing in the fundamentals of a robust, equitable society—one that lives with and respects social groupings and ecosystems in all their depth, diversity, and difference (Dressler et al. 2010).

In conclusion, we hope that this brief outline will help to direct future inquiry into Nature™ Inc. and its alternatives toward engagement with the following questions, among others: What initiatives, existing in the interstices of the current neoliberal order, effectively challenge or subvert this order? How might such initiatives be scaled up in the future? How can they provide productive models for innovative efforts elsewhere? How might new possibilities beyond current practice be conceptualized? What novel theoretical frameworks, or rearticulations of older ones, might offer inspiration for such efforts? How might theory and practice be combined into a productive new praxis? How, finally, might all of this be brought together in pursuit of a world beyond the myopic horizons of neoliberal capitalism?

Bibliography

Abram, D. 1996. *The Spell of the Sensuous: Perception and Language in a More-than-Human World*. London: Vintage Books.

Adams, W. 2004. *Against Extinction: The Story of Conservation*. London: Earthscan.

Adorno, T., and M. Horkheimer. 1986. *Dialectic of Enlightenment*. Translated by J. Cumming. London: Verso.

Agamben, G. 1993. *The Coming Community*. Minneapolis: University of Minnesota Press.

Agrawal, A., and S. Narain. 1991. *Global Warming in an Unequal World: A Case of Environmental Colonialism*. New Delhi: Centre for Science and Environment.

Agrawal, J., and W. A. Kamakura. 1995. "The Economic Worth of Celebrity Endorsers: An Event Study Analysis." *Journal of Marketing* 59(3): 56–62.

Akram-Lodhi, A. H., and C. Kay. 2010. "Surveying the Agrarian Question (Part 1): Unearthing Foundations, Exploring Diversity." *Journal of Peasant Studies* 37(1): 177–202.

Alexander, J., J. McGregor, and T. O. Ranger. 2000. *Violence and Memory: One Hundred Years in the "Dark Forests" of Matebeleland*. Oxford: James Currey.

Alfredsson, E. 2009. "Perspectives on Cost Efficiency and Technological Change." Paper presented at the Swedish Society for Nature Conservation, Stockholm, September 12.

American Carbon Registry. 2011. *Methodology for REDD: Avoiding Planned Deforestation*. Arlington, TX: Winrock International. http://www.americancarbonregistry.org/carbon-accounting/ACR%20Methodology%20for%20REDD%20-%20Avoiding%20Planned%20Deforestation%20v1.0%20April%202011.pdf.

Anderson, K. 2012. *Offsetting (& CDM): A Guarantee for 100 Years or Just a Clever Scam? From a Climate Change Perspective, Is Offsetting Worse than Doing Nothing?* Manchester: University of Manchester. http://www.tyndall.manchester.ac.uk/news/Offsetting-Planet-Under-Pressure-Conf-March-2012.pdf.

Anker, P. 2001. *Imperial Ecology: Environmental Order in the British Empire, 1895–1945*. Cambridge, MA: Harvard University Press.

Arendt, H. (1958) 1998. *The Human Condition*. Chicago: University of Chicago Press.

Arsel, M. 2012. "Between 'Marx and Markets'? The State, the 'Left Turn' and Nature in Ecuador." *Tijdschrift voor economische en sociale geografie* 103(2): 150–63.

Arsel, M., and N. A. Angel. 2012. "'Stating' Nature's Role in Ecuadorian Development: Civil Society and the Yasuní-ITT Initiative." *Journal of Developing Societies* 28(2): 203–27.

Arsel, M., and B. Büscher. 2012. "Nature™ Inc.: Changes and Continuities in Neoliberal Conservation and Environmental Markets." *Development and Change* 43(1): 53–78.

Arthur, W. B. 1999. *Increasing Returns and Path Dependence in the Economy*. Cambridge: Cambridge University Press.

Austin, J. L. 1961. "Performative Utterances." In *Austin, Philosophical Papers*, edited by J. O. Urmson and G. J. Warnock, 233–52. Oxford: Oxford University Press.

Bäckstrand, K. 2004. "Scientisation vs. Civic Expertise in Environmental Governance: Eco-feminist, Eco-modern, and Post-modern Responses." *Environmental Politics* 13(4): 695–714.

Bakker, K. 2004. *An Uncooperative Commodity: Privatizing Water in England and Wales*. Oxford: Oxford University Press.

———. 2007a. "The 'Commons' versus the 'Commodity': Alter-globalization, Anti-privatization and the Human Right to Water in the Global South." *Antipode* 39(3): 430–55.

———. 2007b. "Neoliberalizing Nature? Market Environmentalism in the Water Supply in England and Wales." In *Neoliberal Environments: False Promises and Unnatural Consequences*, edited by N. Heynen, J. McCarthy, S. Prudham, and P. Robbins, 101–14. New York: Routledge.

———. 2009. "Neoliberal Nature, Ecological Fixes, and the Pitfalls of Comparative Research." *Environment and Planning* A 41:1781–87.

Banerjee, B. 2008. "Necrocapitalism." *Organization Studies* 29(12): 1541–63.

Bateson, G., and C. M. Bateson. 2004. *Angels Fear: Towards an Epistemology of the Sacred*. New York: Hampton Press.

BBOP (Business and Biodiversity Offsets Programme). 2009. *Biodiversity Offset Design Handbook*. Washington, DC: Business and Biodiversity Offsets Programme.

———. 2012. "Standard on Biodiversity Offsets." http://bbop.forest-trends.org/guidelines/standard.pdf.

Bebbington, A. 1999. "Capitals and Capabilities: A Framework for Analyzing Peasant Viability, Rural Livelihoods and Poverty." *World Development* 27(12): 2021–44.

———. 2000. "Reencountering Development: Livelihood Transitions and Place Transformations in the Andes." *Annals of the Association of American Geographers* 90:495–520.

Bebbington, A., S. Guggenheim, E. Olson, and M. Woolcock. 2004. "Exploring Social Capital Debates at the World Bank." *Journal of Development Studies* 40(5): 33–64.

Bebbington, A., and T. Perreault. 1999. "Social Capital, Development and Access to Resources in Highland Ecuador." *Economic Geography* 75(4): 395–418.

Becker, G. 1964. *Human Capital*. New York: Columbia University Press.

Beckert, J. 2011. "Imagined Futures: Fictionality in Economic Action." Discussion Paper 11/8. Max-Planck-Institut für Gesellschaftsforschung, Cologne. http://www.mpifg.de/pu/mpifg_dp/dp11–8.pdf.

Bellamy Foster, J. 2000. *Marx's Ecology: Materialism and Nature*. New York: Monthly Review Press.

———. 2002. *Ecology against Capitalism*. New York: Monthly Review Press.

Benjamin, W. 1979. "Paris, Capital of the 19th Century." In *Reflections: Essays, Aphorisms, Autobiographical Writing*, by W. Benjamin, edited by P. Demetz. New York: Harcourt Brace.

Benjaminsen, T. A., and H. Svarstad. 2010. "The Death of an Elephant: Conservation Discourses versus Practices in Africa." *Forum for Development Studies* 37(3): 385–408.

Bennett, J. 2010. *Vibrant Matter: A Political Ecology of Things*. Durham, NC: Duke University Press.

Berkes, F. 2008. *Sacred Ecology: Traditional Ecological Knowledge and Resource Management*. 2nd ed. London: Taylor and Francis.

Biesele, M. 1993. *Women Like Meat: The Folklore and Foraging Ideology of the Kalahari Ju/'hoan*. Bloomington: Indiana University Press.

Bird-David, N. 1992. "Beyond 'The Original Affluent Society': A Culturalist Reformulation." *Current Anthropology* 33(1): 26–34.

———. 1999. "'Animism' Revisited: Personhood, Environment, and Relational Epistemology." *Current Anthropology* 40 (Supplement): S67–S91.

Bird-David, N., and D. Naveh. 2008. "Relational Epistemology, Immediacy, and Conservation: Or, What Do the Nayaka Try to Conserve?" *Journal for the Study of Religion, Nature and Culture* 2(1): 55–73.

Bishop, M., and M. Green. 2008. *Philanthrocapitalism: How the Rich Can Save the World and Why We Should Let Them*. London: A&C Black.

Blackman, A., and R. T. Woodward. 2010. "User Financing in a National Payments for Environmental Services Program: Costa Rican Hydropower." *Ecological Economics* 69:1626–38.

Boas, T., and J. Gans-Moore. 2009. "Neo-liberalism: From New Liberal Philosophy to Anti-liberal Slogan." *Studies in Comparative International Development* 44(1): 137–61.

Böhm, S., and S. Dabhi. 2009. *Upsetting the Offset: The Political Economy of Carbon Markets*. London: Mayfly Books.

Boltanski, L., and E. Chiapello. 2007. *The New Spirit of Capitalism*. London: Verso.

Bond, P. 2010. "Climate Justice Politics across Space and Scale." *Human Geography* 3(2): 49–62.

Bonner, R. 1993. *At the Hand of Man*. London: Simon & Schuster.

Borras, S. 2007. "'Free Market,' Export-Led Development Strategy and Its Impact on Rural Livelihoods, Poverty and Inequality: The Philippine Experience Seen from a Southeast Asian Perspective." *Review of International Political Economy* 14(1): 143–57.

Borras, S., Jr., R. Hall, I. Scoones, B. White, and W. Wolford. 2011. "Towards a Better Understanding of Global Land Grabbing: An Editorial Introduction." *Journal of Peasant Studies* 38(2): 209–16.

Bosso, C. J. 2005. *Environment, Inc.: From Grassroots to Beltway*. Lawrence: University Press of Kansas.

Bourdieu, P. 1977. *Outline of a Theory of Practice*. Cambridge: Cambridge University Press.

———. 1980. "Le capital social: Notes provisoires" [Social capital: Provisional notes]. *Actes de la récherche en sciences sociales* 31:2–3.

———. 1984. *Distinction: A Social Critique of the Judgment of Taste*. Cambridge, MA: Harvard University Press.

———. 1986. "The Forms of Capital." In *Handbook of Theory and Research for the Sociology of Education*, translated by Richard Nice, edited by J. G. Richardson, 231–58. New York: Greenwood Press. [Originally published in 1983 as "Ökonomisches

Kapital, kulturelles Kapital, soziales Kapital." In *Soziale Ungleichheiten*, edited by R. Kreckel, 183–98. Goettingen: Otto Schartz and Co.]

———. 1990. *The Logic of Practice*. Stanford, CA: Stanford University Press.

———. 1994. *The Field of Cultural Production*. Cambridge, MA: Harvard University Press.

———. 1998a. *Acts of Resistance: Against the Tyranny of the Market*. New York: New Press. [Originally published in 1998 as *Contre-feux: Propos pour servir à la résistance contre l'invasion néo-libérale*. Translated by Richard Nice.]

———. 1998b. "The Essence of Neoliberalism." *Le monde diplomatique* (December). http://mondediplo.com/1998/12/08bourdieu.

———. 2003. *Firing Back: Against the Tyranny of the Market 2*. New York: New Press. [Originally published in 2001 as *Contre-feux 2: Pour un mouvement social européen*. Translated by Loïc Wacquant.]

Bourdieu, P., and L. Wacquant. 1992. *An Invitation to Reflexive Sociology*. Chicago: University of Chicago Press.

Boyd, W. 2001. "Making Meat." *Technology and Culture* 42(4): 631–64.

Bracking, S. 2012. "How Do Investors Value Environmental Harm/Care? Private Equity Funds, Development Finance Institutions and the Partial Financialization of Nature-Based Industries." *Development and Change* 43(1): 271–93.

Brain, R. 1993. *Going to the Fair: Readings in the Culture of 19th Century Exhibitions*. Cambridge: Whipple Museum of the History of Science.

Brainard, L., and V. LaFleur. 2007. "Making Poverty History? How Activists, Philanthropists and the Public Are Changing Human Development." Paper presented at the Brookings Blum Roundtable, Washington, DC, 2007.

Brand, U., and C. Görg. 2008. "Post-Fordist Governance of Nature: The Internationalization of the State and the Case of Genetic Resources—a Neo-Poulantzian Perspective." *Review of International Political Economy* 15(4): 567–89.

Braun, B. 2003. "'On the Raggedy Edge of Risk': Articulations of Race and Nature after Biology." In *Race, Nature, and the Politics of Difference*, edited by D. S. Moore, A. Pandian, and J. Kosek, 175–203. Durham, NC: Duke University Press.

Brenner, N., and N. Theodore. 2002. *Spaces of Neoliberalism: Urban Restructuring in Western Europe and North America*. Oxford: Blackwell.

Brenner, N., J. Peck, and N. Theodore. 2010. "After Neoliberalization?" *Globalizations* 7(3): 327–45.

Bridge, G. 2000. "The Social Regulation of Resource Access and Environmental Impact." *Geoforum* 31(2): 237–56.

———. 2008. "Global Production Networks and the Extractive Sector: Governing Resource-Based Development." *Journal of Economic Geography* 8(3): 389–419.

Briggs, B. D. J., D. A. Hill, and R. Gillespie. 2009. "Habitat Banking: How It Could Work in the UK." *Journal for Nature Conservation* 17:112–22.

Broad, R., and J. Cavanagh. 1993. *Plundering Paradise: The Struggle for the Environment in the Philippines*. Berkeley: University of California Press.

Brockington, D. 2002. *Fortress Conservation: The Preservation of the Mkomazi Game Reserve, Tanzania*. Oxford: James Currey.

———. 2008. "Powerful Environmentalisms: Conservation, Celebrity and Capitalism." *Media Culture & Society* 30(4): 551–68.

——— . 2009. *Celebrity and the Environment: Fame, Wealth and Power in Conservation.* London: Zed Books.

——— . 2011. "Ecosystem Services and Fictitious Commodities." *Environmental Conservation* 38(4): 367–69.

——— . 2012. "A Radically Conservative Vision? The Challenge of UNEP's *Towards a Green Economy.*" *Development and Change* 43(1): 409–22.

Brockington, D. 2014. *Celebrity Advocacy and International Development.* London: Routledge.

Brockington, D., and R. Duffy, eds. 2010a. "Capitalism and Conservation." Special issue of *Antipode* 42(3).

——— . 2010b. "Capitalism and Conservation: The Production and Reproduction of Biodiversity Conservation." *Antipode* 42(3): 469–84.

Brockington, D., R. Duffy, and J. Igoe. 2008. *Nature Unbound: Conservation, Capitalism and the Future of Protected Areas.* London: Earthscan.

Brockington, D., and J. Igoe. 2006. "Eviction for Conservation: A Global Overview." *Conservation & Society* 4(3): 424–70.

Brockington, D., and K. Scholfield. 2010. "The Conservationist Mode of Production and Conservation NGOs in Sub-Saharan Africa." *Antipode* 42(3): 551–75.

Brondizio, E., E. Ostrom, and O. Young. 2009. "Connectivity and the Governance of Multilevel Social-Ecological Systems: The Role of Social Capital." *Annual Review of Environment and Resources* 34:253–78.

Brondo, K. V., and N. Brown. 2011. "Neoliberal Conservation, Garifuna Territorial Rights and Resource Management in the Cayos Cochinos Marine Protected Area." *Conservation and Society* 9(2): 91–105.

Brosius, J. P. 2004. "Indigenous Peoples and Protected Areas at the World Parks Congress." *Conservation and Biology* 18:609–12.

——— . 2006. "Seeing Communities: Technologies of Visualization in Conservation." In *Reconsidering Community: The Unintended Consequences of an Intellectual Romance,* edited by Gerald Creed, 227–54. Santa Fe, NM: School of American Research Press.

Brosius, J. P., and L. Campbell. 2010. "Collaborative Event Ethnography: Conservation and Development Trade-Offs at the Fourth World Conservation Congress." *Conservation and Society* 8(4): 245–55.

Buchanan, I., and N. Thoburn. 2008. *Deleuze and Politics.* Edinburgh: Edinburgh University Press.

Buhner, S. H. 2002. *The Lost Language of Plants: The Ecological Importance of Plant Medicines to Life on Earth.* Vermont: Chelsea Green Publishing.

Bumpus, A. 2011. "The Matter of Carbon: Understanding the Materiality of tCO_2e in Carbon Offsets." *Antipode* 43(3): 612–38.

Bumpus, A., and D. Liverman. 2008. "Accumulation by Decarbonisation and the Governance of Carbon Offsets." *Economic Geography* 84(2): 127–55.

Burkett, P. 2005. *Marxism and Ecological Economics: Toward a Red and Green Political Economy.* Brill: Leiden.

Büscher, B. 2008. "Conservation, Neoliberalism and Social Science: A Critical Reflection on the SCB 2007 Annual Meeting, South Africa." *Conservation Biology* 22(2): 229–31.

——— . 2009. "Letters of Gold: Enabling Primitive Accumulation through Neoliberal Conservation." *Human Geography* 2(3): 91–93.

———. 2010a. "Anti-politics as Political Strategy: Neoliberalism and Transfrontier Conservation and Development in Southern Africa." *Development and Change* 41(1): 29–51.

———. 2010b. "Derivative Nature: Interrogating the Value of Conservation in 'Boundless Southern Africa.'" *Third World Quarterly* 31(2): 259–76.

———. 2010c. "Seeking 'Telos' in the 'Transfrontier': Neoliberalism and the Transcending of Community Conservation in Southern Africa." *Environment and Planning A* 42(3): 644–60.

———. 2012. "Payments for Ecosystem Services as Neoliberal Conservation: (Reinterpreting) Evidence from the Maloti-Drakensberg, South Africa." *Conservation and Society* 10(1): 29–41.

Büscher, B., and M. Arsel. 2012. "Introduction: Capitalist Modernity, Neoliberal Conservation and Uneven Geographical Development." *Journal of Economic and Social Geography* 103(2): 129–35.

Büscher, B., and V. Davidov, eds. 2013. *The Ecotourism-Extraction Nexus: Political Economies and Rural Realities of (Un)Comfortable Bedfellows*. London: Routledge.

Büscher, B., and W. Dressler. 2007. "Linking Neoprotectionism and Environmental Governance: On the Rapidly Increasing Tensions between Actors in the Environment-Development Nexus." *Conservation and Society* 5(4): 586–611.

———. 2012. "Commodity Conservation: The Restructuring of Community Conservation in South Africa and the Philippines." *Geoforum* 43(3): 367–76.

Büscher, B., and J. Igoe. 2013. "Prosuming Conservation? Web 2.0, Nature, and the Materiality of Sign Values in Late Capitalism." *Journal of Consumer Culture* 13(3): 283–305.

Büscher B., S. Sullivan, K. Neves, J. Igoe, and D. Brockington. 2012. "Towards a Synthesized Critique of Neoliberal Biodiversity Conservation." *Capitalism, Nature Socialism* 23(2): 4–30.

Büscher, B., and W. Whande. 2007. "Whims of the Winds of Time? Contestations in Biodiversity Conservation and Protected Areas Management." *Conservation and Society* 5(1): 22–43.

Button, J. 2008. "Carbon: Commodity or Currency? The Case for an International Carbon Market Based on the Currency Model." *Harvard Environmental Law Review* 32:571–96.

California Environmental Justice Movement. 2010. EJ Matters website, http://www.ejmatters.org.

Çalışkan, K., and M. Callon. 2009. "Economization, Part 1: Shifting Attention from the Economy towards Processes of Economization." *Economy and Society* 38(3): 369–98.

———. 2010. "Economization, Part 2: A Research Programme for the Study of Markets." *Economy and Society* 39(1): 1–32.

Callon, M. 1998a. "An Essay on Framing and Overflowing: Economic Externalities Revisited by Sociology." In *The Laws of the Markets*, edited by M. Callon, 244–69. Oxford: Blackwell.

———. 1998b. "Introduction: The Embeddedness of Economic Markets in Economics." In *The Laws of the Markets*, edited by M. Callon, 1–57. Oxford: Blackwell.

———. 2005. "Why Virtualism Paves the Way to Political Impotence: A Reply to Daniel Miller's Critique of the Laws of the Markets." *Economic Sociology* 6(2): 3–20.

Callon, M., and K. Çalışkan. 2005. "New and Old Directions in the Anthropology of Markets." Paper prepared for the Wenner-Gren Foundation for Anthropological Research, New York.

Callon, M., and F. Muneisa. 2005. "Economic Markets as Calculative Collective Devices." *Organization Studies* 26(8): 1229–50.

Carney, D. 1998. "Implementing the Sustainable Rural Livelihoods Approach." Paper presented at the DfID Natural Resource Advisors' Conference, Department for International Development, London.

———. 2003. "Sustainable Livelihoods Approaches: Progress and Possibilities for Change." Department for International Development (DfID), London. http://www.eldis.org.

Carolan, M. S. 2005. "Society, Biology, and Ecology: Bringing Nature Back into Sociology's Disciplinary Narrative through Critical Realism." *Organization & Environment* 18:393–421.

Carr, E. S. 2010. "Enactments of Expertise." *Annual Review of Anthropology* 39(1): 17–32.

Carrier, James G. 2010. "Protecting the Environment the Natural Way: Ethical Consumption and Commodity Fetishism." *Antipode* 42(3): 672–89.

Carrier, J. G., and D. V. L. Macleod. 2005. "Bursting the Bubble: The Socio-cultural Context of Ecotourism." *Journal of the Royal Anthropological Institute* 11:315–34.

Carrier, J. G., and D. Miller, eds. 1998. *Virtualism: A New Political Economy*. Oxford: Berg Publishers.

Carrier, J. G., and P. West, eds. 2009. *Virtualism, Governance and Practice: Vision and Execution in Environmental Conservation*. New York: Berghahn.

Carroll, N., J. Fox, and R. Bayon, eds. 2008. *Conservation and Biodiversity Banking: A Guide to Setting Up and Running Biodiversity Credit Trading Systems*. London: Earthscan.

Castells, M. 2000. *The Rise of the Network Society*. Malden, MA: Blackwell.

Castree, N. 2000. "The Production of Nature." In *A Companion to Economic Geography*, edited by E. Sheppard and T. Barnes, 257–69. Oxford: Blackwell.

———. 2003. "Commodifying What Nature?" *Progress in Human Geography* 27(3): 273–97.

———. 2005. *Nature*. London: Routledge.

———. 2006. "A Congress of the World." *Science as Culture* 15(2): 159–70.

———. 2008a. "Neoliberalising Nature: The Logics of Deregulation and Reregulation." *Environment and Planning A* 40(1): 131–52.

———. 2008b. "Neoliberalising Nature: Processes, Effects, and Evaluations." *Environment and Planning A* 40(1): 153–73.

———. 2010a. "Crisis, Continuity and Change: Neoliberalism, the Left and the Future of Capitalism." *Antipode* 41(S1): 185–213.

———. 2010b. "Neoliberalism and the Biophysical Environment: A Synthesis and Evaluation of the Research." *Environment and Society: Advances in Research* 1:5–45.

Castree, N., and B. Braun, eds. 2001. *Social Nature: Theory, Practice, Politics*. Oxford: Wiley-Blackwell.

Chaffin, J. 2010. "Economic Crisis Cuts European Carbon Emissions." *Financial Times*, April 1.

Chaiken, M. 1994. "Economic Strategies and Success on the Philippine Frontier." *Research in Economic Anthropology* 15:277–305.

Chan, K., R. Pringle, J. Ranganathan, C. Briggs, Y. Chan, R. Ehrlich, P. Haff, N. Heller, K. Al-Krafaji, and D. Macmynowski. 2007. "When Agendas Collide: Human Welfare and Biological Conservation." *Conservation Biology* 21(1): 59–68.

Chapin, M. 2004. "A Challenge to Conservationists." *WorldWatch* 17(6): 17–31.

Child, M. 2009. "The Thoreau Ideal as a Unifying Thread in the Conservation Movement." *Conservation Biology* 23(1): 241–43.

Clad, J., and M. Vitug. 1988. "The Politics of Plunder." *Far Eastern Economic Review* 24:48–52.

Clastres, P. (1974) 1989. *Society against the State: Essays in Political Anthropology.* New York: Zone Books.

Cohen, E. 1986. "Law, Folklore and Animal Lore." *Past and Present* 110:6–37.

Coleman, J. 1988. "Social Capital in the Creation of Human Capital." *American Journal of Sociology* 94 (suppl.): S95–120.

Comaroff, J., and J. L. Comaroff. 2002. "Alien-nation: Zombies, Immigrants and Millennial Capitalism." *South Atlantic Quarterly* 101:779–805.

Conelly, T. 1983. "Upland Development in the Tropics: Alternative Economic Strategies in a Philippine Frontier Community." Ph.D. diss., Department of Anthropology, University of California, Santa Barbara.

———. 1985. "Copal and Rattan Collecting in the Philippines." *Economic Botany* 39(1): 39–46.

Connerton, P. 2009. *How Modernity Forgets.* Cambridge: Cambridge University Press.

Coole, D., and S. Frost, eds. 2010. *New Materialisms: Ontology, Agency, and Politics.* Durham, NC: Duke University Press.

Cooper, M. 2010. "Turbulent Worlds: Financial Markets and Environmental Crisis." *Theory, Culture & Society* 27(2–3): 167–90.

Corson, C. 2010. "Shifting Environmental Governance in a Neoliberal World: USAID for Conservation." *Antipode* 42(3): 576–602.

Corson, C., K. I. MacDonald, and B. Neimark, eds. 2013. "Grabbing Green." Special issue of *Human Geography* 6(1).

Costanza, R. 2008. "Natural Capital." In *Encyclopedia of Earth.* Washington, DC: Environmental Information Coalition, National Council for Science and the Environment. http://www.eoearth.org/article/Natural_capital, accessed June 15, 2011.

Costanza, R., and H. Daly. 1992. "Natural Capital and Sustainable Development." *Conservation Biology* 6(1): 37–46.

Costanza, R., R. d'Arge, S. de Groot, M. Farber, B. Grasso, K. Hannon, S. Limburg, R. Naeem, J. O'Neill, R. Paruelo, R. Raskin, P. Sutton, and M. van den Belt. 1997. "The Value of the World's Ecosystem Services and Natural Capital." *Nature* 387:253–60.

Cramb, R. A., C. J. P. Colfer, W. Dressler, P. Laungaramsri, M. E. Le Quang Trung, N. L. Peluso, and R. L. Wadley. 2009. "Swidden Transformations and Rural Livelihoods in Southeast Asia." *Human Ecology* 37(3): 323–46.

Crilly, R. 2010. *Saving Darfur: Everyone's Favourite African War.* London: Reportage Press.

Cronon, W. 1991. *Nature's Metropolis: Chicago and the Great West.* New York: W. W. Norton.

———. 1996. "The Trouble with Wilderness." In *Uncommon Ground: Rethinking the Human Place in Nature*, edited by W. Cronon, 69–90. New York: W. W. Norton.

Crory, J. 2002. "Spectacle, Attention, and Counter-Memory." In *Guy Debord and the Situationist International*, edited by T. McDonough, 455–66. Cambridge, MA: MIT Press.

Crouch, C. 2004. *Post-Democracy*. Cambridge: Polity Press.

———. 2011. *The Strange Non-death of Neoliberalism*. Cambridge: Polity Press.

Curry, P. 2008. "Nature Post-nature." *New Formations* 26(Spring): 51–64.

———. 2011. *Ecological Ethics: An Introduction*. Cambridge: Polity Press.

Czech, B. 2013. *Supply Shock: Economic Growth at the Crossroads and the Steady State Solution*. Gabriola Island: New Society Publishers.

Daily, G. C. 1997. *Nature's Services: Societal Dependence on Natural Ecosystems*. Washington, DC: Island Press.

Daily, G. C., and K. Ellison. 2002. *The New Economy of Nature: The Quest to Make Conservation Profitable*. Washington, DC: Island Press.

Darrier, E. 1996. "Environmental Governmentality: The Case of Canada's Green Plan." *Environmental Politics* 5:585–606.

Dean, J. 2008. "Enjoying Neoliberalism." *Cultural Politics* 4(1): 47–72.

Debord, G. (1967) 1995. *Society of the Spectacle*. London: Rebel Press.

———. (1971) 2008. *A Sick Planet*. London: Seagull Books.

———. (1988) 1998. *Comments on Society of the Spectacle*. London: Verso Books.

DEFRA. 2012. *Biodiversity Offsetting Pilots Technical Paper: The Metric for the Biodiversity Offsetting Pilot in England*. London: DEFRA. http://www.defra.gov.uk/publications/2012/04/02/pb13745-bio-tech-paper/.

Deleuze, G. 2001. *Pure Immanence: Essays on a Life*. Translated by A. Boyman. New York: Zone Books.

Deleuze, G., and F. Guattari. (1972) 2004. *Anti-Oedipus: Capitalism and Schizophrenia*. Translated by R. Hurley, M. Seem, and H. R. Lane. London: Continuum.

———. (1980) 1987. *A Thousand Plateaus: Capitalism and Schizophrenia*. Translated by B. Massumi. London: Athlone Press.

Dempsey, J. 2011. "Making Markets, Making Biodiversity: Understanding Global Biodiversity Politics." Ph.D. diss., University of British Columbia.

———. 2013. "Biodiversity Loss as Material Risk: Tracking the Changing Meanings and Materialities of Biodiversity Conservation." *Geoforum* 45:41–51.

Descartes, R. (1637) 1968. *Discourse on Method*. London: Penguin Books.

Descola, P. 2005. *Par-delá nature et culture*. Paris: Gallimard.

Descola, P., and G. Pálsson. 1996a. Introduction to *Nature and Society: Anthropological Perspectives*, edited by Philippe Descola and G. Pálsson, 1–21. London: Routledge.

———, eds. 1996b. *Nature and Society: Anthropological Perspectives*. London: Routledge.

Deutsche Bank. 2009. *The Long and Short of It: Power Sector Key to EUA and CER Prices*. Carbon Emissions Commodities Report. London: Deutsche Bank, May 5.

De Vries, P. 2007. "Don't Compromise Your Desire for Development! A Lacanian/Deleuzian Rethinking of the Anti-politics Machine." *Third World Quarterly* 28(1): 25–43.

Diamond, J., and M. Gilpin. 1983. "Biographical Value (Umbilkin) and the Origin of the Philippine Avifauna." *Oikos* 41:301–41.

Dietz, R., and D. O'Neill. 2013. *Enough Is Enough: Building a Sustainable Economy in a World of Finite Resources*. San Francisco: Berret-Koehler.

Dobson, A. 2010. "Democracy and Nature: Speaking and Listening." *Political Studies* 58(4): 752–68.

Docena, H. 2010. *The CDM in the Philippines: Costly, Dirty, Money-Making Schemes*. Bangkok: Focus on the Global South.

——. 2011. "Guilt, Blame, and Innocence in the International Climate Change Negotiations: The (Im)moral Origins of the Global Carbon Market." Unpublished paper, University of California, Berkeley, September.

Dore, D. 1999. "Economic Assessment of Alternative Settlement Options for Residents in the Gwayi and Bembesi State Forest Reserves." Background paper prepared for the Shared Forest Management Project/Department for International Development and Forestry Commission, Forestry Commission, Harare.

Douglas, M. 1966. *Purity and Danger.* London: Routledge and Kegan Paul.

Dowie, M. 2009. *Conservation Refugees: The Hundred-Year Conflict between Global Conservation and Native Peoples.* Boston, MA: MIT Press.

Dressler, D., B. Büscher, M. Schoon, D. Brockington, T. Hayes, C. Kull, J. McCarthy, and K. Streshta. 2010. "From Hope to Crisis and Back? A Critical History of the Global CBNRM Narrative." *Environmental Conservation* 37(1): 5–15.

Dressler, W. 2006. "Co-opting Conservation: Migrant Resource Control and Access to National Park Management in the Philippine Uplands." *Development and Change* 37(2): 401–26.

——. 2009. *Old Thoughts in New Ideas: State Conservation Measures, Livelihood and Development on Palawan Island.* Quezon City: Ateneo de Manila University Press.

——. 2011. "First to Third Nature: The Rise of Capitalist Conservation in Palawan, the Philippines." *Journal of Peasant Studies* 38(3): 533–57.

Dressler, W., and R. Roth. 2010. "The Good, the Bad, and the Contradictory: Neoliberal Conservation Governance in Rural Southeast Asia." *World Development* 39(5): 851–62.

Dreyfus, H., and P. Rabinow. 1983. *Michel Foucault: Beyond Structuralism and Hermeneutics.* Chicago: University of Chicago Press.

Driesen, D. 2003. *The Economic Dynamics of Environmental Law.* Cambridge, MA: MIT Press.

Duffy, R. 2002. *Trip Too Far: Ecotourism, Politics, and Exploitation.* London: Earthscan.

——. 2006. "The Potentials and Pitfalls of Global Environmental Governance: The Politics of Transfrontier Conservation Areas in Southern Africa." *Political Geography* 25:89–112.

——. 2008. "Neoliberalising Nature: Global Networks and Ecotourism Development in Madagascar." *Journal of Sustainable Tourism* 16(3): 327–44.

——. 2010. *Nature Crime. How We're Getting Conservation Wrong.* New Haven, CT: Yale University Press.

——. 2012. "The International Political Economy of Tourism and the Neoliberalisation of Nature: Challenges Posed by Selling Close Interactions with Animals." *Review of International Political Economy* 20(3): 605–26.

Duffy, R., and L. Moore. 2010. "Neoliberalising Nature? Elephant-Back Tourism in Thailand and Botswana." *Antipode* 42(3): 742–66.

Duggan, L. 2012. *The Twilight of Equality: Neoliberalism, Cultural Politics and the Attack on Democracy.* Boston, MA: Beacon Press.

Dyer, N., and S. Counsell. 2010. *McREDD: How McKinsey "Cost-Curves" Are Distorting REDD.* London: Rainforest Foundation.

Dyer-Witheford, N. 1999. *Cyber-Marx: Cycles and Circuits of Struggle in High-Technology Capitalism.* Urbana: University of Illinois Press.

Eagleton, T. 1990. *The Ideology of the Aesthetic.* Oxford: Blackwell.

Eder, J. 1987. *On the Road to Tribal Extinction*. Berkeley: University of California Press.

———. 1999. *A Generation Later: Household Strategies and Economic Change in the Rural Philippines*. Quezon City: Ateneo de Manila University Press.

———. 2006. "Land Use and Economic Change in the Post-frontier Upland Philippines." *Land Degradation and Development* 17:149–58.

Eder, J., and J. Fernandez. 1996. "Palawan, a Last Frontier." In *Palawan at the Crossroads: Development and the Environment on a Philippine Frontier*, edited by J. Eder and J. Fernandez, 1–23. Quezon City: Ateneo de Manila University Press.

Ehrenfeld, D. 2008. "Neoliberalization of Conservation." *Conservation Biology* 22(5): 1091–92.

———. 2009. *Becoming Good Ancestors: How We Balance Nature, Community, and Technology*. Oxford: Oxford University Press.

Ehrlich, P. R., and A. H. Ehrlich. 1981. *Extinction: The Causes and Consequences of the Disappearance of Species*. New York: Random House.

EIA (Environmental Investigation Agency). 2010. "Q&A on Industrial Gases and the CDM in the EU ETS." EIA, London. http://www.eia-international.org/files/reports212–1.pdf.

ELDIS. 2011. "Livelihoods Assets." Institute of Development Studies, Sussex University. http://www.eldis.org/go/topics/dossiers/livelihoods-connect/what-are-livelihoods-approaches/livelihoods-assets, accessed June 17, 2011.

Emerson, R. W. (1836) 1985. *Nature*. London: Penguin Books.

Engel, S., S. Pagiola, and S. Wunder. 2008. "Designing Payments for Environmental Services in Theory and Practice: An Overview of the Issues." *Ecological Economics* 65:663–74.

Escobar, A. 1995. *Encountering Development: The Making and Unmaking of the Third World*. Princeton, NJ: Princeton University Press.

Fairhead, J., M. Leach, and I. Scoones. 2012a. "Green Grabbing." Special issue of *Journal of Peasant Studies* 39(2).

———. 2012b. "Green Grabbing: A New Appropriation of Nature?" *Journal of Peasant Studies* 39(2): 237–61.

Farrington, J., D. Carney, C. Ashley, and C. Turton. 1999. "Sustainable Livelihoods in Practice: Early Applications of Concepts in Rural Areas." Overseas Development Institute. *Natural Resource Perspectives* 42:1–15. http://www.odi.org.uk.

FASE. 2003. "Open Letter to Executives and Investors in the Prototype Carbon Fund." FASE, Espiritu Santo, May 23.

Federici, S. 2004. *Caliban and the Witch: Women, the Body and Primitive Accumulation in Medieval Europe*. New York: Autonomedia.

Ferguson, J. 1994. *The Anti-politics Machine: "Development," Depoliticization, and Bureaucratic Power in Lesotho*. Minneapolis: University of Minnesota Press.

———. 2006. *Global Shadows: Africa in the Neo-liberal World*. Durham, NC: Duke University Press.

———. 2010. "The Uses of Neoliberalism." *Antipode* 41(1): 166–84.

Ferris, K. O. 2007. "The Sociology of Celebrity." *Sociology Compass* 1(1): 371–84.

Fine, B. 2001. *Social Capital versus Social Theory: Political Economy and Social Science at the Turn of the Millennium*. London: Routledge.

———. 2010. *Theories of Social Capital: Researchers Behaving Badly*. New York: Pluto.

Fink, B. 1995. *The Lacanian Subject: Between Language and Jouissance*. Princeton, NJ: Princeton University Press.

Fletcher, R. 2007. "Free Play: Transcendence as Liberation." In *Beyond Resistance: The Future of Freedom*, edited by R. Fletcher, 143–62. New York: Nova Science Publishers.

——. 2009. "Ecotourism Discourse: Challenging the Stakeholder Theory." *Journal of Ecotourism* 8(3): 269–85.

——. 2010. "Neoliberal Environmentality: Towards a Poststructuralist Political Ecology of the Conservation Debate." *Conservation and Society* 8(3): 171–81.

——. 2011. "Sustaining Tourism, Sustaining Capitalism? The Tourism Industry's Role in Global Capitalist Expansion." *Tourism Geographies* 13(3): 443–61.

——. 2012a. "The Art of Forgetting: Imperialist Amnesia and Public Secrecy." *Third World Quarterly* 33(3): 447–63.

——. 2012b. "Capitalizing on Chaos: Climate Change and Disaster Capitalism." *Ephemera* 12(1/2): 97–112.

——. 2012c. "Using the Master's Tools? Neoliberal Conservation and the Evasion of Inequality." *Development and Change* 43(1): 295–317.

——. 2013. "Bodies Do Matter: The Peculiar Persistence of Neoliberalism in Environmental Governance." *Human Geography* 6(1): 29–45.

Fletcher, R., and J. Breitling. 2012. "Market Mechanism or Subsidy in Disguise? Governing Payment for Environmental Services in Costa Rica." *Geoforum* 43(3): 402–11.

Fletcher, R., and K. Neves. 2012. "Contradictions in Tourism: The Promise and Pitfalls of Ecotourism as a Manifold Capitalist Fix." *Environment and Society: Advances in Research* 3(1): 60–77.

Flew, T. 2011. "Michel Foucault's *The Birth of Biopolitics* and Contemporary Neoliberalism Debates." *Thesis Eleven* 108(1): 44–65.

Flint, J., and A. De Waal. 2008. *Darfur: A New History of a Long War*. London: Zed Books.

Foreman, J. 2009. "How Hollywood Finds Its Charitable Causes." *Sunday Times*, October 4.

Forsyth, T. 2003. *Critical Political Ecology: The Politics of Environmental Science*. New York: Routledge.

Foucault, M. (1966) 1970. *The Order of Things: An Archaeology of the Human Sciences*. London: Routledge.

——. (1976) 1998. *The Will to Knowledge: The History of Sexuality, Volume 1*. Translated by R. Hurley. London: Penguin Books.

——. 1982. "The Subject and Power." In *Michel Foucault: Beyond Structuralism and Hermeneutics*, edited by H. Dreyfus and P. Rabinow. Chicago: University of Chicago Press.

——. 1991. *Remarks on Marx: Conversations with Duccio Trombadori*. New York: Semiotext(e).

——. 2007. *Security, Territory, and Population: Lectures at the Collège de France, 1977–1978*. New York: Palgrave.

——. 2008. *The Birth of Biopolitics: Lectures at the Collège de France 1978–1979*. Translated by G. Burchell. Basingstoke: Palgrave Macmillan.

Fox, R. 1954. "Tagbanua Religion and Society." Ph.D. diss., University of Chicago and the National Museum, Manila.

Franke, A., ed. 2012a. "Animism." *e-flux* 36. http://www.e-flux.com/issues/36-july-2012/.

———. 2012b. "Animism: Notes on an Exhibition." *e-flux* 36. http://www.e-flux.com/journal/animism-notes-on-an-exhibition/.

Freud, S. 1962. *Civilization and Its Discontents*. New York: Norton.

Frischmann, B., and M. Lemley. 2006. "Spillovers." *Columbia Law Review* 100:101–46.

Gabler, N. 1998. *Life: The Movie. How Entertainment Conquered Reality*. New York: Vintage Books.

Gamson, J. 2000. "The Web of Celebrity." *American Prospect*, September, 40–41.

Garcia-Parpet, M. 2008. "The Social Construction of a Perfect Market: The Strawberry Auction at Fontaines-en-Sologne." In *Do Economists Make Markets? On the Performativity of Economics*, edited by D. MacKenzie, F. Muniesa, and L. Siu, 22–50. Princeton, NJ: Princeton University Press.

Garland, E. 2008. "The Elephant in the Room: Confronting the Colonial Character of Wildlife Conservation in Africa." *African Studies Review* 51(3): 51–74.

Garuba, H. 2012. "On Animism, Modernity/Colonialism, and the African Order of Knowledge: Provisional Reflections." *e-flux*. http://www.e-flux.com/journal/on-animism-modernitycolonialism-and-the-african-order-of-knowledge-provisional-reflections/.

Geertz, C. 1973. *The Interpretation of Cultures*. New York: Basic.

Gibson-Graham, J. K. 2006. *A Postcapitalist Politics*. Minneapolis: University of Minnesota Press.

———. 2011. "A Feminist Project of Belonging for the Anthropocene." *Gender, Place, and Culture* 18(11): 1–21.

Gibson-Graham, J. K., J. Cameron, and S. Healy. 2013. *Take Back the Economy: An Ethical Guide for Transforming Our Communities*. Minneapolis: University of Minnesota Press.

Giddens, A. 1984. *The Constitution of Society: An Outline of the Theory of Structuration*. London: Polity Press.

Gilbert, J. 2004. "Becoming-Music: The Rhizomatic Movement of Improvisation." In *Deleuze and Music*, edited by Ian Buchanan and S. Swiboda, 118–38. Edinburgh: Edinburgh University Press.

Gillenwater, M. 2012. *Getting Real about "Real" Carbon Offsets*. Washington, DC: Greenhouse Gas Institute. http://ghginstitute.org/2012/08/03/getting-real-about-real-carbon-offsets/.

Gilmore, G. W. 1919. *Animism or Thought Currents of Primitive Peoples*. London: Kessinger Publishing.

Ginsberg, A. 1956. "Footnote to Howl." In *Howl and Other Poems*. San Francisco: City Light Books.

Glynos, J. 2012. "The Place of Fantasy in a Critical Political Economy: The Case of Market Boundaries." *Cardozo Law Review* 33(6): 2373–2411.

Goeminne, G. 2012. "Lost in Translation: Climate Denial and the Return of the Political." *Global Environmental Politics* 12(2): 1–8.

Goldman, M. 2005. *Imperial Nature: The World Bank and Struggles for Social Justice in the Age of Globalization*. New Haven, CT: Yale University Press.

Goldman, M., P. Nadasdy, and M. D. Turner. 2011. *Knowing Nature: Conservations at the Intersection of Political Ecology and Science Studies*. Chicago: University of Chicago Press.

Goldman, R. 1994. "Contradictions in a Political Economy of Sign Value." *Current Perspectives in Social Theory* 14:183–211.

Goldman, R., and S. Papson. 2006. "Capital's Brandscapes." *Journal of Consumer Culture* 6:327–53.

——. 2011. *Landscapes of Capital*. Cambridge: Polity Press.

Goldstein, R. 2012. "An Ecology of the Self: When Nature Goes Public and Other Wild Thoughts." Paper presented at the Rappaport panel, American Anthropological Association conference, San Francisco, November 14–18, 2012.

Goodland, R., and J. Anhang. 2010. "Livestock and Climate Change." *Worldwatch*, November/December, 10–19.

Gotham, K. 2009. "Creating Liquidity out of Spatial Fixity: The Secondary Circuit of Capital and the Subprime Mortgage Crisis." *International Journal of Urban and Regional Research* 33:355–71.

Graeber, D. 2011. *Debt: The First Five Thousand Years*. Brooklyn, NY: Melville House.

Graham, P. 2007. *Hypercapitalism: New Media, Language, and Social Perceptions of Value*. New York: Peter Lang.

Gramsci, A. 1971. *Selections from the Prison Notebooks*. Edited and translated by Q. Hoare and G. Nowell Smith. New York: International Publishers.

Gray, J. 2002. *Straw Dogs: Thoughts on Humans and Other Animals*. London: Granta Books.

Green, N. 1990. *The Spectacle of Nature: Landscape and Bourgeois Culture in 19th-Century France*. Manchester: Manchester University Press.

Greenwood, R., and R. Suddaby. 2006. "Institutional Entrepreneurship in Mature Fields: The Big Five Accounting Firms." *Academy of Management Journal* 49(1): 27–48.

Gregersen, H., H. El Lakany, A. Karsenty, and A. White. 2010. *Does the Opportunity Cost Approach Indicate the Real Cost of REDD+? Rights and Realities of Paying for REDD+*. Washington, DC: Rights and Resources Initiative.

Grove, R. H. 1995. *Green Imperialism: Colonial Expansion, Tropical Island Edens and the Origins of Environmentalism, 1600–1860*. Cambridge: Cambridge University Press.

Grugel, J., and P. Riggirozzi. 2012. "Post-neoliberalism in Latin America: Rebuilding and Reclaiming the State after Crisis." *Development and Change* 43(1): 1–21.

Guattari, F. (1989) 2000. *The Three Ecologies*. Translated by I. Pindar and P. Sutton. London: Continuum.

Guha, R. 1989. *The Unquiet Woods: Ecological Change and Villager Resistance in the Himalaya*. New Delhi: Oxford University Press.

Guthman, J., and M. DuPuis. 2006. "Embodying Neoliberalism: Economy, Culture and the Politics of Fat." *Environment and Planning D: Society and Space* 24:427–48.

Haber, A. 2012. "Severo's Severity and Antolín's Paradox." e-flux 36. http://www.e-flux.com/journal/severo%E2%80%99s-severity-and-antolin%E2%80%99s-paradox/.

Hall, S. 1986. "On Postmodernism and Articulation: An Interview with Stuart Hall." *Journal of Communication Inquiry* 10(2): 45–60.

Hall, S., D. Massey, and M. Rustin. 2013. *After Neoliberalism: Analysing the Present*. http://www.lwbooks.co.uk/journals/soundings/manifesto.html.

Hamilton, R. 2011. *Fighting for Darfur: Public Action and the Struggle to Stop Genocide*. New York: Palgrave Macmillan.

Hanchard, M. G. 2006. "A Theory of Quotidian Politics." In *Party/Politics: Horizons in Black Political Thought*, 25–67. Oxford: Oxford University Press.

Hannis, M., and S. Sullivan. 2012. *Offsetting Nature? Habitat Banking and Biodiversity Offsets in the English Land Use Planning System*. Dorset: Green House. http://www .greenhousethinktank.org/files/greenhouse/home/Offsetting_nature_inner_final.pdf.

Haraway, D. 2008. *When Species Meet*. Minneapolis: University of Minnesota Press.

Harcourt, W. 2012. *Women Reclaiming Sustainable Livelihoods: Spaces Lost, Spaces Gained*. London: Palgrave.

Hardie, I., and D. MacKenzie. 2007. "Assembling an Economic Actor: The Agencement of a Hedge Fund." *Sociological Review* 55(1): 57–80.

Harris, A. P. 2008. "The Wisdom of the Body: Embodied Knowing in Eco-paganism." Ph.D. diss., University of Winchester.

Harrison, P. 1992. "Descartes on Animals." *Philosophical Quarterly* 42(169): 219–27.

Harriss, J. 2002. *Depoliticising Development: The World Bank and Social Capital*. London: Anthem.

Harvey, D. 1989. *The Condition of Postmodernity: An Inquiry into the Origins of Cultural Change*. Oxford: Basil Blackwell.

———. 2000. *Spaces of Hope*. Berkeley: University of California Press.

———. 2003. *The New Imperialism*. Oxford: Oxford University Press.

———. 2005. *A Brief History of Neoliberalism*. Oxford: Oxford University Press.

———. 2006a. *The Limits to Capital*. 2nd ed. London: Verso.

———. 2006b. *Spaces of Global Capitalism: A Theory of Uneven Geographical Development*. 2nd ed. London: Verso.

———. 2010. *The Enigma of Capital and the Crises of Capitalism*. Oxford: Oxford University Press.

Harvey, G. 2005. *Animism: Respecting the Living World*. London: Hurst and Co.

Haynes, D., and G. Prakash. 1991. "Introduction: The Entanglement of Power and Resistance." In *Contesting Power: Resistance and Everyday Social Relations in South Asia*, edited by D. Haynes and G. Prakash. New Delhi: Oxford University Press.

Henderson, G. L. 2003. *California and the Fictions of Capital*. Philadelphia: Temple University Press.

Hepburn, R. W. 1984. *Wonder, and Other Essays*. Edinburgh: Edinburgh University Press.

Heynen, N., J. McCarthy, P. Robbins, and S. Prudham, eds. 2007. *Neoliberal Environments: False Promises and Unnatural Consequences*. New York: Routledge.

Heynen, N., and P. Robbins. 2005. "The Neoliberalization of Nature: Governance, Privatization, Enclosure and Valuation." *Capitalism Nature Socialism* 16(1): 5–8.

Hildyard, N. 2008. *A (Crumbling) Wall of Money: Financial Bricolage, Derivatives and Power*. Dorset: Cornerhouse. http://www.thecornerhouse.org.uk/resource/ crumbling-wall-money.

Hodgson, N. J. 1989. "Preliminary Report on the Tenant Farmer Programme in the Ngamo, Gwaai and Bembesi Forest Areas." Unpublished paper prepared for the Ministry of Lands, Agriculture and Rural Resettlement, Harare.

Hoff, A. 1997. "The Water Snake of the Khoekhoen and |Xam." *South African Archaeological Bulletin* 52:21–37.

Holert, T. 2012. "'A Live Monster That Is Fruitful and Multiplies': Capitalism as Poisoned Rat?" e-flux 36. http://www.e-flux.com/journal/%E2%80%9Ca-live-monster -that-is-fruitful-and-multiplies%E2%80%9D-capitalism-as-poisoned-rat/.

Hollander, J. A., and R. L. Einwohner. 2004. "Conceptualizing Resistance." *Sociological Forum* 194:533–52.

Holm, P. 2001. "The Invisible Revolution: The Construction of Institutional Change in the Fisheries." Ph.D. diss., Norwegian College of Fishery, Tromsø.

———. 2008. "Which Way Is Up on Callon?" In *Do Economists Make Markets? On the Performativity of Economics*, edited by D. MacKenzie, F. Muniesa, and L. Siu, 225–43. Princeton, NJ: Princeton University Press.

Holmes, G. 2007. "Protection, Protest and Politics: Understanding Resistance to Conservation." *Conservation and Society* 5(2): 184–201.

———. 2010. "The Rich, the Powerful and the Endangered: Conservation Elites, Networks and the Dominican Republic." *Antipode* 42(3): 624–46.

Honey, M. 2008. *Ecotourism and Sustainable Development: Who Owns Paradise?* 2nd ed. New York: Island Press.

Hornborg, A. 2006. "Animism, Fetishism, and Objectivism as Strategies for Knowing (or Not Knowing) the World." *Ethnos* 71(1): 21–32.

Hughes, D. 2005. "Third Nature: Making Space and Time in the Great Limpopo Conservation Area." *Cultural Anthropology* 20(2): 157–84.

Huws, U. 2011. "Crisis as Capitalist Opportunity: New Accumulation through Public Service Commodification." In *Socialist Register 2012: The Crisis and the Left*, edited by L. Panitch, G. Albo, and V. Chibber, 64–84. New York: Monthly Review Press.

IFAD. 2011. "The Sustainable Livelihoods Approach." The International Fund for Agricultural Development. http://www.ifad.org/sla/index.htm.

Igoe, J. 2004. *Conservation and Globalization: A Study of Indigenous Communities and National Parks from East Africa to South Dakota*. Belmont, CA: Wadsworth/Thompson.

———. 2010. "The Spectacle of Nature in the Global Economy of Appearances: Anthropological Engagements with the Spectacular Mediations of Transnational Biodiversity Conservation." *Critique of Anthropology* 30(4): 375–97.

———. 2013a. "Consume, Connect, Conserve: Consumer Spectacle and the Technical Mediation of Neoliberal Conservation's Aesthetic of Redemption and Repair." *Human Geography* 6(1): 16–28.

———. 2013b. "Nature on the Move II: Making, Marketing, and Managing an Accessible and Penetrable Nature That Seems to Dominate Our Environment by Virtue of Its Circulation." Unpublished paper.

Igoe, J., and D. Brockington. 2007. "Neoliberal Conservation: A Brief Introduction." *Conservation and Society* 5(4): 432–49.

Igoe, J., and B. Croucher. 2007. "Conservation, Commerce, and Communities: The Story of Community-Based Wildlife Management in Tanzania's Northern Tourist Circuit." *Conservation and Society* 5(4): 534–61.

Igoe, J., K. Neves, and D. Brockington. 2010. "A Spectacular Eco-Tour around the Historic Bloc: Theorising the Convergence of Biodiversity Conservation and Capitalist Expansion." *Antipode* 42(3): 486–512.

IMF (International Monetary Fund). 2009. *2009 Annual Report*. Geneva: IMF.

Ingold, T. 2000. *The Perception of the Environment: Essays in Livelihood, Dwelling and Skill*. London: Routledge.

———. 2005. "Epilogue: Towards a Politics of Dwelling." *Conservation and Society* 3(2): 501–8.

————. 2006. "Rethinking the Animate, Re-animating Thought." *Ethnos* 71(1): 9–20.

————. 2010. "Bringing Things to Life: Creative Entanglements in a World of Materials." ESRC National Centre for Research Methods, NCRM Working Paper Series 05/10.

————. 2011. *Being Alive*. London: Routledge.

Ingold, T., and E. Hallam, eds. 2007. *Creativity and Cultural Improvisation*. Oxford: Berg.

IPCC (Intergovernmental Panel on Climate Change). 1996. *Climate Change 1995: The Science of Climate Change*. Cambridge: Cambridge University Press.

Irigaray, L. 1997. "The Other: Woman." In *Feminisms*, edited by S. Kemp and J. Squires, 308–15. Oxford: Oxford University Press.

IUCN (International Union for Conservation of Nature). 2012. "World Conservation Congress: Congress Themes Explained." http://iucn.org/knowledge/focus/2012_world_conservation_congress/interviews/.

Jacoby, K. 2001. *Crimes against Nature: Squatters, Poachers, Thieves and the Hidden History of American Conservation*. Berkeley: University of California Press.

Jameson, F. 1991. *Post-modernism, or, The Cultural Logic of Late Capitalism*. Durham, NC: Duke University Press.

————. 2003. "Future City." *New Left Review* 21:65–79.

Jansson, A., M. Hammer, C. Folke, and R. Costanza, eds. 1994. *Investing in Natural Capital: The Ecological Economics Approach to Sustainability*. Washington, DC: Island Press.

Jappe, A. 1999. *Guy Debord*. Berkeley: University of California Press.

Jones, A. 1999. "Dialectics and Difference: Against Harvey's Dialectical Post-Marxism." *Progress in Human Geography* 23(4): 529–55.

Jones, C. 2013. *Can The Market Speak?* London: Zero Books.

Kallis, G. 2011. "In Defence of Degrowth." *Ecological Economics* 70(5): 873–80.

Kapoor, I. 2005. "Participatory Development, Complicity and Desire." *Third World Quarterly* 26(8): 1203–20.

Karpik, L. 2010. *Valuing the Unique: The Economics of Singularities*. Princeton, NJ: Princeton University Press.

Kerkvliet, B. 1974. "Land Reform in the Philippines since the Marcos Coup." *Pacific Affairs* 47(3): 286–304.

Keynes, J. M. 1936. *The General Theory of Interest, Employment and Money*. London: Macmillan.

||Khumub, K., A. Botelle, R. Scott, and C. Bosman. 2007. *Journey of a Rain Shaman*. Windhoek: Mamokobo Research and Productions.

Kiernan, M. J. 2009. *Investing in a Sustainable World: Why Green Is the New Color of Money on Wall Street*. New York: Amacom.

Kingsbury, P. 2010. "Locating the Melody of the Drives." *Professional Geographer* 62(1): 519–33.

————. 2011. "Sociospatial Sublimation: The Human Resources of Love in Sandals Resorts International, Jamaica." *Annals of the Association of American Geographers* 101(3): 650–69.

Kloppenburg, J. 1988. *First the Seed: The Political Economy of Plant Biotechnology*. Cambridge: Cambridge University Press.

Kosoy, N., and E. Corbera. 2010. "Payments for Ecosystem Services as Commodity Fetishism." *Ecological Economics* 69:1228–36.

Kovel, J. 2002. *The Enemy of Nature: The End of Capitalism or the End of the World?* London: Zed Books.

Kress, J. 1977. "Contemporary and Prehistoric Subsistence Patterns on Palawan." In *Cultural Ecological Perspectives on Southeast Asia*, edited by W. Wood, 29–47. Athens: Ohio State University Press.

Kuper, A. 1993. "The Return of the Native." *Current Anthropology* 44(3): 389–95.

Kwashirai, V. C. 2009. *Green Colonialism: 1890–1980*. New York: Cambria Press.

Kysar, D. 2010. "Not Carbon Offsets, but Carbon Upsets." *Guardian*, August 29.

Lampel, J., and A. Meyer. 2008. "Field-Configuring Events as Structuring Mechanisms." *Journal of Management Studies* 45(6): 1025–35.

Landell-Mills, N., and I. Porras. 2002. *Silver Bullet or Fools' Gold? A Global Review of Markets for Forest Environmental Services and Their Impact on the Poor*. London: IIED.

Lang, C. 2011. "McKinsey's Advice on REDD Is 'Fundamentally Flawed,' Says Greenpeace." *REDD Monitor*, April 8. http://www.redd-monitor.org/2011/04/08/mckinsey-advice-on-redd-is-fundamentally-flawed-says-greenpeace/.

Lansing, A. 1959. *Endurance: Shackleton's Incredible Voyage*. New York: Carroll and Graf.

Latour, B. 1987. *Science in Action: How to Follow Scientists and Engineers through Society*. Cambridge, MA: Harvard University Press.

———. 1993. *We Have Never Been Modern*. Cambridge, MA: Harvard University Press.

———. 2004. *Politics of Nature: How to Bring the Sciences into Democracy*. Cambridge, MA: Harvard University Press.

———. 2005. *Reassembling the Social: An Introduction to Actor-Network-Theory*. Oxford: Oxford University Press.

———. 2010a. "An Attempt at a 'Compositionist Manifesto.'" *New Literary History* 41:471–90.

———. 2010b. *On the Modern Cult of the Factish Gods*. Durham, NC: Duke University Press.

Layton, L. 2009. "Irrational Exuberance: Neoliberal Subjectivity and the Perversion of Truth." *Subjectivity* 3(3): 303–22.

Leach, M., J. Fairhead, and J. Fraser. 2012. "Green Grabs and Biochar: Revaluing African Soils and Farming in the New Carbon Economy." *Journal of Peasant Studies* 39(2): 285–307.

Lee, B., and E. LiPuma. 2002. "Cultures of Circulation: The Imaginations of Modernity." *Public Culture* 14:191–213.

Le Guin, U. K. 2000. *The Telling*. New York: Ace Books.

Lekan, T. 2011. "Serengeti Shall Not Die! Bernhard Grzimek, Wildlife Film, and the Makings of a Tourist Landscape in Tanzania." *German History* 29(2): 224–64.

Lemke, T. 2001. "'The Birth of Bio-politics': Michel Foucault's Lecture at the Collège de France on Neo-liberal Governmentality." *Economy and Society* 30(2): 190–207.

Letcher, A. 2003. "'Gaia Told Me to Do It': Resistance and the Idea of Nature within Contemporary British Eco-paganism." *Ecotheology* 8(1): 61–84.

Levine, A. 2002. "Convergence or Convenience? International Conservation NGOs and Development Assistance in Tanzania." *World Development* 30(6): 1043–55.

Lewis, J. 2008. "*Ekila*: Blood, Bodies, and Egalitarian Societies." *Journal of the Royal Anthropological Institute* 14:297–315.

———. 2008/9. "Managing Abundance, Not Chasing Scarcity: The Real Challenge for the 21st Century." *Radical Anthropology* 2:11–18.

Lewis-Williams, J. D., and D. G. Pearce. 2004. *San Spirituality: Roots, Expression, and Social Consequences.* New York: Altamira Press.

Leyshon, A., and N. Thrift. 2007. "The Capitalisation of Almost Everything: The Future of Finance and Capitalism." *Theory, Culture & Society* 24:97–115.

Li, T. M. 2007. *The Will to Improve: Governmentality, Development, and the Practice of Politics.* Durham, NC: Duke University Press.

LiPuma, E., and B. Lee. 2004. *Financial Derivatives and the Globalization of Risk.* Durham, NC: Duke University Press.

———. 2005. "Financial Derivatives and the Rise of Circulation." *Economy and Society* 34:404–27.

Lohmann, L. 2006. *Carbon Trading: A Critical Conversation on Climate, Privatization and Power.* Uppsala: Dag Hammarskjöld Foundation.

———. 2009. "Toward a Different Debate in Environmental Accounting: The Cases of Carbon and Cost-Benefit." *Accounting, Organizations and Society* 34(3–4): 499–534.

———. 2011. "The Endless Algebra of Climate Markets." *Capitalism Nature Socialism* 22(4): 93–116.

———. 2012. "Financialization, Commodification and Carbon: The Contradictions of Neoliberal Climate Policy." *Socialist Register* 48:85–107.

Lopez, M. 1987. "The Politics of Land at Risk in a Philippine Frontier." In *Lands at Risk in the Third World: Local Level Perspectives*, edited by P. D. Little and M. Horowitz, 230–48. Boulder, CO: Westview Press.

Lorimer, J. 2010. "Elephants as Companion Species: The Lively Biogeographies of Asian Elephant Conservation in Sri Lanka." *Transactions of the Institute of British Geographers* 35(4): 491–506.

———. 2012. "Multinatural Geographies for the Anthropocene." *Progress in Human Geography* 36(5): 593–612.

Low, C. 2008. *Khoisan Medicine in History and Practice.* Cologne: Rüdiger Köppe Verlag.

Low, C., and S. Sullivan. Forthcoming. "Shades of the Rainbow Serpent? A Khoesān Animal between Myth and Landscape in Southern Africa—Ethnographic Contextualisations of Rock Art Representations." *Arts*, special issue on African rock art.

Lukács, G. 1971. *History and Class Consciousness: Studies in Marxist Dialectics.* Cambridge, MA: MIT Press.

Luke, T. 1999. "Environmentality as Green Governmentality." In *Discourses of the Environment*, edited by E. Darrier, 121–51. Malden, PA: Blackwell Publishers.

MacDonald, C. 2008. *Green, Inc.: An Environmental Insider Reveals How a Good Cause Has Gone Bad.* Guilford: Lyons Press.

———. 2009. "A Good Cause Gone Bad." *Adbusters* #81. https://www.adbusters.org/magazine/81/environment.html.

MacDonald, K. 2010a. "Business, Biodiversity and New 'Fields' of Conservation: The World Conservation Congress and the Renegotiation of Organizational Order." *Conservation and Society* 8(4): 256–75.

———. 2010b. "The Devil Is in the (Bio)diversity: Private Sector 'Engagement' and the Restructuring of Biodiversity Conservation." *Antipode* 42(3): 513–50.

MacDonald, K. I., and C. Corson. 2012. "'TEEB Begins Now': A Virtual Moment in the Production of Natural Capital." *Development and Change* 43(1): 159–84.

MacKenzie, D. 2003. "An Equation and Its Worlds: Bricolage, Exemplars, Disunity and Performativity in Financial Economics." *Social Studies of Science* 33(6): 831–68.

———. 2006. "Making Things the Same: Gases, Emission Rights and the Politics of Carbon Markets." *Accounting, Organizations and Society* 34(3–4): 440–55.

———. 2009. *An Engine, Not a Camera: How Financial Models Shape Markets*. Cambridge, MA: MIT Press.

MacKenzie, D., and Y. Millo. 2003. "Constructing a Market, Performing Theory: The Historical Sociology of a Financial Derivatives Exchange." *American Journal of Sociology* 109(1): 107–45.

MacKenzie, D., F. Muniesa, and L. Siu, eds. 2007. *Do Economists Make Markets? On the Performativity of Economics*. Princeton, NJ: Princeton University Press.

MacKenzie, J. M. 1997. *The Empire of Nature: Hunting, Conservation and British Imperialism*. Manchester: Manchester University Press.

Magdoff, F., and J. Bellamy Foster. 2011. *What Every Environmentalist Needs to Know about Capitalism: A Citizen's Guide to Capitalism and the Environment*. New York: Monthly Review Press.

Mandel, J., J. Donlan, and J. Armstrong. 2010. "A Derivative Approach to Endangered Species Conservation." *Frontiers in Ecology and the Environment* 8(1): 44–49.

Mansfield, B. 2004. "Neoliberalism in the Oceans: Rationalization, Property Rights and the Commons Question." *Geoforum* 35(3): 131–26.

Mapedza, E. 2007. "Forestry Policy in Colonial and Postcolonial Zimbabwe: Continuity and Change." *Journal of Historical Geography* 33:833–51.

Marazzi, C. 2011. *The Violence of Financial Capitalism*. Los Angeles: Semiotext(e).

Marcus, G. E. 1995. "Ethnography in/of the World System: The Emergence of Multi-sited Ethnography." *Annual Review of Anthropology* 24:95–117.

———. 2000. *Para-sites: A Casebook against Cynical Reason*. Chicago: University of Chicago Press.

Marshall, D. P. 1997. *Celebrity and Power: Fame in Contemporary Culture*. Minneapolis: University of Minnesota Press.

Martinez-Alier, J. 2003. *The Environmentalism of the Poor: A Study of Ecological Conflicts and Valuation*. Cheltenham: Edward Elgar.

Marx, K. (1853) 1962. "The British Rule in India." Reprinted from the *New York Times Tribune*, June 25, 1853. In *The First Indian War of Independence: 1857–1859*, by Karl Marx and Frederick Engels, 13–19. Moscow: Progress Publishers.

———. (1858) 1973. *Grundrisse: Foundations of the Critique of Political Economy (Rough Draft)*. Translated by Martin Nicolaus. London: Penguin Classics.

———. (1867) 1976. *Capital. Volume I*. London: Penguin Books.

———. (1870) 1978. *Capital. Volume II*. London: Penguin Books.

———. 1975. *Collected Works*. Translated by R. Dixon et al. New York: International Publishers.

Matose, F. 2002. "Local People and Reserved Forests in Zimbabwe: What Prospects for Co-management?" Ph.D. diss., University of Sussex, Brighton, UK.

———. 2008. "Institutional Configurations around Forest Reserves in Zimbabwe: The Challenge of Nested Institutions for Resource Management." *Local Environment: The International Journal of Justice and Sustainability* 13(5): 1–12.

———. Forthcoming. "Forestry Discourses on Zimbabwe's Protected Forests 1900–2000." *Journal of Southern African Studies*.

McAfee, K. 1999. "Selling Nature to Save It? Biodiversity and Green Developmental-
ism." *Environment and Planning D: Society and Space* 17(1): 133–54.
———. 2012a. "The Contradictory Logic of Global Ecosystem Services Markets."
Development and Change 43(1): 105–31.
———. 2012b. "Nature in the Market-World: Ecosystem Services and Inequality."
Development 55(1): 25–33.
McAfee, K., and E. N. Shapiro. 2010. "Payments for Ecosystem Services in Mexico:
Nature, Neoliberalism, Social Movements, and the State." *Annals of the Association
of American Geographers* 100(3): 579–99.
McCarthy, J. 2012. "The Financial Crisis and Environmental Governance 'after' Neo-
liberalism." *Tijdschrift voor economische en social geografie* 103(2): 180–95.
McCarthy, J., and S. Prudham. 2004. "Neoliberal Nature and the Nature of Neoliber-
alism." *Geoforum* 35(1): 275–83.
McDermott, M. 2000. "Boundaries and Pathways: Indigenous Identities, Ancestral
Domain, and Forest Use in Palawan, the Philippines." Ph.D. diss., University of
California, Berkeley.
McGregor, J. 1991. "Woodland Resources: Ecology, Policy, and Ideology—a Historical
Case Study of Woodland Use in Shurugwi Communal Area, Zimbabwe." Ph.D.
diss., Loughborough University of Technology, Loughborough.
McMillan, C. 2008. "Symptomatic Readings: Žižekian Theory as a Discursive Strategy."
International Journal of Žižek Studies 2(1): 1–22.
McShane, T. O., P. D. Hirsch, T. C. Trung, A. N. Songorwa, A. Kinzig, B. Monteferri,
D. Mutekanga, H. V. Thang, J. L. Dammert, M. Pulgar-Vidal, M. Welch-Devine,
J. P. Brosius, P. Coppolillo, and S. O'Connor. 2011. "Hard Choices: Making Trade-
Offs between Biodiversity Conservation and Human Well-Being." *Biological Con-
servation* 144(3): 966–72.
Meadows, D. H., D. L. Meadows, and J. Randers. 1972. *The Limits to Growth*. New
York: Universe Books.
Meiksins Wood, E. 2002. *The Origin of Capitalism: A Longer View*. London: Verso.
Melitopoulos, A., and M. Lazzarato. 2012a. "Assemblages: Félix Guattari and Machinic
Animism." *e-flux* 36. http://www.e-flux.com/journal/assemblages-felix-guattari-and
-machinic-animism/.
———. 2012b. "Machinic Animism." *Deleuze Studies* 6:240–49.
Mellor, M. 2000. "Feminism and Environmental Ethics: A Materialist Perspective."
Ethics and the Environment 5(1): 107–23.
Merchant, C. (1980) 1989. *The Death of Nature: Women, Ecology and the Scientific
Revolution*. New York: Harper and Row.
Merrifield, A. 2011. *Magical Marxism: Subversive Politics and the Imagination*. London:
Pluto Press.
Meyer, D. S., and J. Gamson. 1995. "The Challenge of Cultural Elites: Celebrities and
Social Movements." *Sociological Inquiry* 65(2): 181–206.
Mhuriro, L. 1996. "Resettlement Impact on Gwayi/Bembesi Forest Areas Ecology and
Functioning." Unpublished paper prepared for the Forestry Commission, Bulawayo.
Midgley, M. (2004) 2011. *The Myths We Live By*. London: Routledge.
Millennium Ecosystem Assessment. 2005. *Ecosystems and Human Well-Being*: Synthesis.
Washington, DC: Island Press.
Miller, D. 2005. "A Reply to Callon." *Economic Sociology* 6(3): 3–13.

Milne, S., and B. Adams. 2012. "Market Masquerades: Uncovering the Politics of Community-Level Payments for Environmental Services in Cambodia." *Development and Change* 43:133–58.

Mirowski, P. 2011. *Science-Mart: Privatizing American Science*. Cambridge, MA: Harvard University Press.

Mitchell, T. 2008. "Rethinking Economy." *Geoforum* 39(1): 1116–21.

Mitman, G. 1999. *Reel Nature: America's Romance with Wildlife Film*. Cambridge, MA: Harvard University Press.

Moeller, N. 2010. "The Protection of Traditional Knowledge in the Ecuadorian Amazon: A Critical Ethnography of Capital Expansion." Ph.D. diss., University of Lancaster.

Monfreda, C. 2010. "Setting the Stage for New Global Knowledge: Science, Economics, and Indigenous Knowledge in 'The Economics of Ecosystems and Biodiversity' at the Fourth World Conservation Congress." *Conservation and Society* 8(4): 276–85.

Monod, J. 1972. *Chance and Necessity*. Translated by A. Wainhouse. Glasgow: Collins.

Moore, D. S. 2005. *Suffering for Territory: Race, Place, and Power in Zimbabwe*. Durham, NC: Duke University Press.

Moore, J. 2010. "The End of the Road? Agricultural Revolutions in the Capitalist World-Ecology, 1450–2010." *Journal of Agrarian Change* 10(3): 389–413.

Morris, A. W. 2006. "Easing Conservation? Conservation Easements, Public Accountability and Neoliberalism." *Geoforum* 39:1215–27.

Mosse, D. 2006. "Anti-social Anthropology? Objectivity, Objection, and the Ethnography of Public Policy and Professional Communities." *Journal of the Royal Anthropological Institute* 12(1): 935–56.

——. 2010. "A Relational Approach to Durable Poverty, Inequality and Power." *Journal of Development Studies* 46(7): 1156–78.

Munden Group. 2011. *REDD and Forest Carbon: Market-Based Critique and Recommendations*. New York: Munden Group.

Mushove, P. T. 1993. "Can Wildlife Utilization Alone Sustain Forest Operations on Gwayi/Mbembesi Forests?" In *The Ecology and Management of Indigenous Forests in Southern Africa: Proceedings of an International Symposium, Victoria Falls, Zimbabwe, 27–29 July 1992*, edited by G. D. Pearce and D. J. Gumbo, 330–35. Harare: Forestry Commission and SAREC.

Naidoo, R. L., C. Weaver, M. de Longchamp, and P. du Plessis. 2011. "Namibia's Community-Based Natural Resource Management Programme: An Unrecognized Payments for Ecosystem Services Scheme." *Environmental Conservation* 38:445–53.

Nancy, J.-L. 2001. "The Two Secrets of the Fetish." *Diacritics* 31(2): 3–8.

Natural Capital Project. 2012. "Natural Capital Project: Aligning Economic Forces with Conservation." http://www.naturalcapitalproject.org.

Nealon, J. 2008. *Foucault beyond Foucault: Power and Its Intensification since 1984*. Stanford, CA: Stanford University Press.

Neimark, B. 2012. "Industrializing Nature, Knowledge, and Labour: The Political Economy of Bioprospecting in Madagascar." *Geoforum* 43(5): 980–90.

Neumann, R. P. 1998. *Imposing Wilderness: Struggles over Livelihood and Nature Preservation in Africa*. Berkeley: University of California Press.

Neves, K. 2006. "Politics of Environmentalism and Ecological Knowledge at the Intersection of Local and Global Processes." *Journal of Ecological Anthropology* 10:19–32.

———. 2009a. "The Sacredness of Human-Cetacean Unities: Towards a Non-mechanical Non-transcendental Ethnographic Approach." Paper presented at the American Academy of Religion Conference, Montreal, Canada, November 7–10.

———. 2009b. "Urban Botanical Gardens and the Aesthetics of Ecological Learning: A Theoretical Discussion and Preliminary Insights from Montreal's Botanical Garden." *Anthropologica* 51:145–57.

———. 2010. "Cashing in on Cetourism: A Critical Ecological Engagement with Dominant E-NGO Discourses on Whaling, Cetacean Conservation, and Whale Watching." *Antipode* 42:719–41.

———. N.d. "The Politics of Multi-faceted World Heritage: Pico Vineyard's Cultural Landscape." Unpublished manuscript.

Nevins, J., and N. Peluso. 2008. *Taking Southeast Asia to Market: Commodities, Nature, and People in the Neoliberal Age.* Ithaca, NY: Cornell University Press.

Newell, P., and M. Paterson. 2010. *Climate Capitalism: Global Warming and the Transformation of the Global Economy.* Cambridge: Cambridge University Press.

Norgaard, R. B. 2010. "Ecosystem Services: From Eye-Opening Metaphor to Complexity Blinder." *Ecological Economics* 69(6): 1219–27.

Novellino, D., and W. Dressler. 2010. "The Role of 'Hybrid' NGOs in the Conservation and Development of Palawan Island, the Philippines." *Society and Natural Resources* 23(2): 165–80.

Oates, J. F. 1999. *Myth and Reality in the Rain Forest: How Conservation Strategies Are Failing in West Africa.* Berkeley: University of California Press.

Oberzaucher, E., and K. Grammer. 2008. "Everything Is Movement: On the Nature of Embodied Communication." In *Embodied Communication,* edited by I. Wachsmuth and G. Knoblich, 151–77. Oxford: Oxford University Press.

Ocampo, N. 1996. "A History of Palawan." In *Palawan at a Crossroads: Development and the Environment on a Philippine Frontier,* edited by J. Eder and J. Fernandez, 23–37. Quezon City: Ateneo de Manila University Press.

O'Connor, J. 1988. "Capitalism, Nature, and Socialism: A Theoretical Introduction." *Capitalism Nature Socialism* 1:11–38.

———. 1998. *Natural Causes: Essays in Ecological Marxism.* New York: Guilford.

O'Connor, M. 1994a. "On the Misadventures of Capitalist Nature." In *Is Capitalism Sustainable?,* edited by M. O'Connor, 125–51. New York: Guilford Press.

———, ed. 1994b. *Is Capitalism Sustainable? Political Economy and the Politics of Ecology.* New York: Guilford Press.

O'Neill, J. 2006. *Markets Deliberation and Environment.* New York: Routledge.

Ortner, S. 2006. *Social Theory and Anthropology: Culture, Power, and the Acting Subject.* Durham, NC: Duke University Press.

Osterweil, M. 2004. "Place-Based Globalists: Rethinking the *Global* in the Alternative Globalization Movement." Unpublished paper, Department of Anthropology, University of North Carolina, Chapel Hill.

Ostrom, E. 1990. *Governing the Commons: The Evolution of Institutions for Collective Action.* Cambridge: Cambridge University Press.

———. 1994. "Constituting Social Capital and Collective Action." *Journal of Theoretical Politics* 6(4): 527–62.

Ostrom, E., R. Gardner, and J. Walker. 1994. *Rules, Games, and Common-Pool Resources.* Ann Arbor: University of Michigan Press.

Ostrom, E., L. Schroeder, and S. Wynne. 1993. *Institutional Incentives and Sustainable Development*. Boulder, CO: Westview.

Osuoka, I. 2009. "Paying the Polluter? The Relegation of Local Community Concerns in 'Carbon Credit' Proposals of Oil Corporations in Nigeria." Unpublished manuscript.

Ouroussoff, A. 2010. *Wall Street at War*. London: Polity.

Overbeek, H., and B. Van Apeldoorn. 2012. *Neoliberalism in Crisis*. Basingstoke: Palgrave.

Pagiola, S., J. Bishop, and N. Landell-Mills. 2002. *Selling Forest Environmental Services: Market-Based Mechanism for Conservation and Development*. London: Earthscan.

Panelli, R. 2010. "More-than-Human Social Geographies: Posthuman and Other Possibilities." *Progress in Human Geography* 34(1): 79–87.

Park Visitor Statistics. 2009. *Puerto Princesa Subterranean River National Park, City Government of Puerto Princesa City*. Palawan Island.

Pawliczek, J., and S. Sullivan. 2011. "Conservation and Concealment in SpeciesBanking.com, US: An Analysis of Neoliberal Performance in the Species Offsetting Service Industry." *Environmental Conservation* 38(4): 435–44.

Pearce, D. W., A. Markandya, and E. Barbier. 1989. *Blueprint for a Green Economy*. London: Earthscan.

Pearce, F. 2010. "Carbon Trading Tempts Firms to Make Greenhouse Gas." *New Scientist*, December 16.

Peck, J. 2010. *Constructions of Neoliberal Reason*. Oxford: Oxford University Press.

Peck, J., and N. Theodore. 2012. "Reanimating Neoliberalism: Process Geographies of Neoliberalisation." *Social Anthropology* 20(2): 177–85.

Peck, J., and A. Tickell. 2002. "Neoliberalizing Space." *Antipode* 34:380–404.

Peluso, N. L. 1992. *Rich Forests, Poor People: Resource Control and Resistance in Java*. Berkeley: University of California Press.

———. 1993. "Coercing Conservation: The Politics of State Resource Control." *Global Environmental Change* 3(2): 199–217.

———. 2012. "What's Nature Got to Do with It? A Situated Historical Perspective on Socio-natural Commodities." *Development and Change* 43(1): 79–104.

Perdan, S., and A. Azapagic. 2011. "Carbon Trading: Current Schemes and Future Developments." *Energy Policy* 39(10): 6040–54.

Perelman, M. 2007. "Primitive Accumulation from Feudalism to Neoliberalism." *Capitalism Nature Socialism* 18:44–61.

Peters, M. 2008. "The Global Failure of Neoliberalism: Privatize Profits; Socialize Losses." *Global-e* 6 (November). http://global-ejournal.org/2008/11/06/the-global-failure-of-neoliberalism-privatize-profits-socialize-losses/.

Peterson, M., D. Hall, A. Feldpausch-Parker, and T. Peterson. 2009. "Obscuring Ecosystem Function with Application of the Ecosystem Services Concept." *Conservation Biology* 24(1): 113–19.

Philippine Census. 2000. *Census of the Philippine Nation*. Quezon City, the Philippines.

Plant, Sadie. 1998. *Zeros and Ones: Digital Women and the New Technoculture*. London: Fourth Estate.

Plows, A. 1998. "Earth First! Defending Mother Earth Direct-Style." In *DIY Culture: Party and Protest in Nineties Britain*, edited by G. Mckay, 152–73. London: Verso.

Plumwood, V. 2006. "The Concept of a Cultural Landscape: Nature, Culture and Agency in the Land." *Ethics and the Environment* 11(2): 115–50.

Polanyi, K. 1944. *The Great Transformation: The Political and Economic Origins of Our Time*. Boston: Beacon.

Poniewozik, J. 2005. "The Year of Charitainment." *Time*, December 19.

Porritt, J. 2007. *Capitalism as if the World Mattered*. London: Earthscan.

Posey, Darrell A. 2002. *Kayapo Ethnoecology and Culture*. New York: Routledge.

Potter, L. 2009. "Oil Palm and Resistance in West Kalimantan, Indonesia." In *Agrarian Angst and Rural Resistance in Contemporary Southeast Asia*, edited by D. Caouette and S. Turner, 105–34. London: Routledge.

Power, C., and I. Watts. 1997. "The Woman with the Zebra's Penis: Gender, Mutability and Performance." *Journal of the Anthropological Institute* 3(3): 537–60.

Pringle, H. 2004. *Celebrity Sells*. Chichester: John Wiley and Sons.

Putnam, R. 1993. *Making Democracy Work: Civic Traditions in Modern Italy*. Princeton, NJ: Princeton University Press.

Radin, M. 1996. *Contested Commodities*. Cambridge, MA: Harvard University Press.

Randalls, S. 2010. "Weather Profits: Weather Derivatives and the Commercialization of Meteorology." *Social Studies of Science* 40:705–30.

Read, J. 2003. *The Micro-politics of Capital: Marx and the Prehistory of the Present*. Albany, NY: State University of New York Press.

Redford, K. H., and W. M. Adams. 2009. "Payment for Ecosystem Services and the Challenge of Saving Nature." *Conservation Biology* 23(4): 785–87.

Rigg, J., and S. Nattapoolwat. 2001. "Embracing the Global in Thailand: Activism and Pragmatism in an Era of Deagrarianization." *World Development* 29(6): 945–60.

Rival, L. 1996. "Blowpipes and Spears: The Social Significance of Huaorani Technological Choices." In *Nature and Society: Anthropological Perspectives*, edited by Philippe Descola and G. Pálsson, 145–64. London: Routledge.

Robbins, P., and A. Luginbuhl. 2005. "The Last Enclosure: Resisting Privatization of Wildlife in the Western United States." *Capitalism Nature Socialism* 16(1): 45–61.

Robertson, M. M. 2000. "No Net Loss: Wetland Restoration and the Incomplete Capitalization of Nature." *Antipode* 32(4): 463–93.

———. 2004. "The Neoliberalization of Ecosystem Services: Wetland Mitigation Banking and Problems in Environmental Governance." *Geoforum* 35(3): 361–73.

———. 2006. "The Nature That Capital Can See: Science, State, and Market in the Commodification of Ecosystem Services." *Environment and Planning D: Society and Space* 24(3): 367–87.

———. 2007. "Discovering Price in All the Wrong Places: The Work of Commodity Definition and Price under Neoliberal Environmental Policy." *Antipode* 39(3): 500–526.

———. 2010. "Performing Environmental Governance." *Geoforum* 41(1): 7–10.

———. 2012. "Measurement and Alienation: Making a World of Ecosystem Services." *Transactions of the Institute of British Geographers* 37(3): 386–401.

Robertson, M. M., and N. Hayden. 2008. "Evaluation of a Market in Wetland Credits: Entrepreneurial Wetland Banking in Chicago." *Conservation Biology* 22(3): 636–46.

Rojek, C. 2001. *Celebrity*. London: Reaktion Books Ltd.

Rosales, J. 2006. "Economic Growth and Biodiversity Loss in an Age of Tradable Permits." *Conservation Biology* 20(4): 1042–50.

Rose, N., P. O'Malley, and M. Valverde. 2006. "Governmentality." *Annual Review of Law and Social Science* 2(1): 83–104.

Roszak, T. (1992) 2001. *The Voice of the Earth: An Exploration of Ecopsychology.* Grand Rapids, MI: Phanes Press.

Roth, R., and W. Dressler. 2012. "Market-Oriented Conservation Governance: The Particularities of Place." *Geoforum* 43(3): 363–66.

Ruddick, S. 2008. "Towards a Dialectics of the Positive." *Environment and Planning A* 40(11): 2588–2602.

Rushkoff, D. 2011. *Life Inc.: How Corporatism Conquered the World, and How We Can Take It Back.* New York: Random House.

Rutherford, S. 2007. "Green Governmentality: Insights and Opportunities in the Study of Nature's Rule." *Progress in Human Geography* 31(3): 291–307.

Sachedina, Hassanali T. 2008. "Wildlife Is Our Oil: Conservation, Livelihoods and NGOs in the Tarangire Ecosystem, Tanzania." Ph.D. diss., St. Anthony's College.

Sahlins, M. 1974. *Stone Age Economics.* Chicago: Aldine Transaction.

Sand, P. 2010. "The Chagos Archipelago: Footprint of Empire, or World Heritage?" *Environmental Policy and Law* 40(5): 232–42.

Sandor, R. L. 2012. *Good Derivatives: A Story of Financial and Environmental Innovation.* Hoboken, NJ: John Wiley and Sons.

Sawyer, S. 2004. *Crude Chronicles: Indigenous Politics, Multinational Oil, and Neoliberalism in Ecuador.* Durham, NC: Duke University Press.

Sayer, J. 2009. "Reconciling Conservation and Development: Are Landscapes the Answer?" *Biotropica* 41(6): 649–52.

Schmidt, A. 1971. *The Concept of Nature in Marx.* London: NLB.

Schmidt, S. 1998. "Mythical Snakes in Namibia." In *The Proceedings of the Khoisan Identities and Cultural Heritage Conference, Cape Town 12–16 July 1997,* edited by A. Bank, H. Heese, and C. Loff, 269–80. Bellville: Institute for Historical Research, University of the Western Cape and Infosource.

Schmidt-Soltau, K., and D. Brockington. 2007. "Protected Areas and Resettlement: What Scope for Voluntary Relocation." *World Development* 35:2182–2202.

Schneider, L. 2011. "Perverse Incentives under the CDM: An Evaluation of HFC-23 Destruction Projects." *Climate Policy* 11(2): 851–64.

Schwartzman, S. 2010. "Nature and Culture in Central Brazil: Panará Natural Resource Concepts and Tropical Forest Conservation." *Journal of Sustainable Forestry* 29(2): 302–27.

Scoones, I. 1998. "Sustainable Rural Livelihoods: A Framework for Analysis." IDS Working Paper 72, Brighton.

———. 2009. "Livelihoods Perspectives and Rural Development." *Journal of Peasant Studies* 36(1): 171–96.

Scott, J. C. 1985. *Weapons of the Weak: Everyday Forms of Villager Resistance.* New Haven, CT: Yale University Press.

———. 1990. *Domination and the Art of Resistance: Hidden Transcripts.* New Haven, CT: Yale University Press.

———. 1998. *Seeing like a State: How Certain Schemes to Improve the Human Condition Have Failed.* New Haven, CT: Yale University Press.

Scott, W. R., M. Rueff, P. Mendel, and C. A. Caronna. 2000. *Institutional Change and Healthcare Organizations.* Chicago: University of Chicago Press.

Scott Cato, M. 2012. *The Bioregional Economy: Land, Liberty and the Pursuit of Happiness.* London: Routledge.

Seagle, C. 2012. "Inverting the Impacts: Mining, Conservation and Sustainability Claims near the Rio Tinto/QMM Ilmenite Mine in Southeast Madagascar." *Journal of Peasant Studies* 39(2): 447–77.

Sewell, W. 2005. *Logics of History: Social Theory and Social Transformation.* Chicago: University of Chicago Press.

Sikor, T., and P. Tuong Vi. 2005. "The Dynamics of Commoditization in a Vietnamese Upland Village." *Journal of Agrarian Change* 5(3): 405–28.

Singh, N. 2013. "The Affective Labor of Growing Forests and the Becoming of Environmental Subjects: Rethinking Environmentality in Odisha, India." *Geoforum* 47:189–98.

Sireau, N. 2008. *Make Poverty History: Political Communication in Action.* London: Palgrave Macmillan.

Sivaramakrishnan, K. 1999. *Modern Forests: Statemaking and Environmental Change in Eastern Colonial India.* Stanford, CA: Stanford University Press.

———. 2005. "In Focus: Moral Economies, State Spaces, and Categorical Violence: Anthropological Engagements with the Work of James Scott." *American Anthropologist* 107 S: 3.

Sklair, L. 2001. *The Transnational Capitalist Class.* Oxford: Blackwell.

Smith, N. 1996. "The Production of Nature." In *FutureNatural*, edited by G. Robertson, M. Mash, L. Tickner, J. Bird, B. Curtis, and T. Putnam, 35–54. London: Routledge.

———. 2007. "Nature as Accumulation Strategy." In *Coming to Terms with Nature: Socialist Register 2007*, edited by L. Panitch and C. Leys, 16–36. London: Merlin Press.

———. 2008. *Uneven Development: Nature, Capital and the Production of Space.* 3rd ed. Athens: University of Georgia Press.

Snodgrass, J. G., and K. Tiedje. 2008. "Guest Editors' Introduction: Indigenous Nature Reverence and Conservation: Seven Ways of Transcending and Unnecessary Dichotomy." *Journal for the Study of Religion, Nature and Culture* 2(1): 6–29.

Sommerlad, N. 2009. "Charity Balls Up: Do Celebrity Fund-Raisers Short-Change the Causes They Boast about Backing?" *Daily Mirror.* http://blogs.mirror.co.uk/investigations/2009/2006/charity-balls-up-do-celebrity.html.

Soros, G. 2008. *The New Paradigm for Financial Markets: The Credit Crisis of 2008 and What It Means.* New York: Public Affairs.

Spence, M. 1999. *Dispossessing the Wilderness: Indian Removal and the Making of National Parks.* Oxford: Oxford University Press.

Spinoza, B. (1677) 1996. *Ethics.* Translated by E. Curley. London: Penguin.

Springer, S. 2012. "Neoliberalism as Discourse: Between Foucauldian Political Economy and Marxian Poststructuralism." *Critical Discourse Studies* 9(2): 133–47.

Springer, S., A. Ince, J. Pickerill, G. Brown, and A. J. Barker. 2012. "Anarchist Geographies." Special issue of *Antipode* 44(5).

Stavrakakis, Y. 1997a. "Green Fantasy and the Real of Nature: Elements of a Lacanian Critique of Green Ideological Discourse." *Journal for the Psychoanalysis of Culture & Society* 2(1): 123–32.

———. 1997b. "Green Ideology: A Discursive Reading." *Journal of Political Ideologies* 2(3): 259–79.

Steger, M. B., and R. K. Roy. 2010. *Neoliberalism: A Very Short Introduction.* Oxford: Oxford University Press.

Stengers, I. 2012. "Reclaiming Animism." *e-flux* 36. http://www.e-flux.com/journal/reclaiming-animism/.

Stiglitz, J. E. 2008a. "The End of Neoliberalism?" *Project Syndicate*, July 7. http://www.project-syndicate.org/commentary/stiglitz101.

———. 2008b. "Is There a Post-Washington Consensus Consensus?" In *The Washington Consensus Reconsidered*, edited by N. Serra and J. Stiglitz, 41–56. New York: Oxford University Press.

Strathern, M. 2000. *Audit Cultures: Anthropological Studies in Accountability, Ethics and the Academy*. New York: Routledge.

Sukhdev, P. 2010. Preface to *The Economics of Ecosystems and Biodiversity: Ecological and Economic Foundations*, edited by P. Kumar, xvii–xxvii. London: Earthscan.

———. 2012. *Corporation 2020: Transforming Business for Tomorrow's World*. Washington, DC: Island Press.

Sullivan, S. 2006a. "The Elephant in the Room? Problematizing 'New' (Neoliberal) Biodiversity Conservation." *Forum for Development Studies* 33(1): 105–35.

———. 2006b. "On Dance and Difference: Bodies, Movement and Experience in Khoesān Trance-Dancing." In *Talking about People: Readings in Contemporary Cultural Anthropology*, 4th ed., edited by W. A. Haviland, R. Gordon, and L. Vivanco, 234–41. New York: McGraw-Hill.

———. 2009. "Green Capitalism, and the Cultural Poverty of Constructing Nature as Service Provider." *Radical Anthropology* 3:18–27.

———. 2010a. "'Ecosystem Service Commodities'—a New Imperial Ecology? Implications for Animist Immanent Ecologies, with Deleuze and Guattari." *New Formations: A Journal of Culture/Theory/Politics* 69:111–28, special issue on "Imperial Ecologies."

———. 2010b. "The Environmentality of 'Earth Incorporated': On Contemporary Primitive Accumulation and the Financialisation of Environmental Conservation." Paper presented at the conference "An Environmental History of Neoliberalism," May 6–8, http://www.worldecologyresearch.org/.

———. 2011a. "Banking Nature? The Financialisation of Environmental Conservation." Working Papers Series #8, Open Anthropology Cooperative Press. http://openanthcoop.net/press/2011/03/11/banking-nature/.

———. 2011b. "Conservation Is Sexy! What Makes This So, and What Does This Make? An Engagement with *Celebrity and the Environment*." *Conservation and Society* 9(4): 334–45.

———. 2011c. "Supposing Truth Is a Woman? A Commentary." *International Journal of Feminist Politics* 13(2): 231–37.

———. 2012a. "After the Green Rush? Biodiversity Offsets, Uranium Power, and the Calculus of Casualties in Greening Growth." Paper presented to the Annual Meeting of the American Association of Geographers, New York City, February.

———. 2012b. "Financialisation, Biodiversity Conservation and Equity: Some Currents and Concerns." Third World Network, Environment and Development Series 16, Penang, Malaysia. http://twnside.org.sg/title/end/pdf/end16.pdf.

———. 2013a. "After the Green Rush? Biodiversity Offsets, Uranium Power and the 'Calculus of Casualties' in Greening Growth." *Human Geography* 6(1): 80–101.

———. 2013b. "Banking Nature? The Spectacular Financialisation of Environmental Conservation." *Antipode* 45(1): 198–217.

———. 2013c. "The Natural Capital Myth." Public Political Ecology Lab. http://ppel.arizona.edu/blog/2013/03/15/natural-capital-myth.

Sullivan, S., and R. Rohde. 2002. "On Non-equilibrium in Arid and Semi-arid Grazing Systems." *Journal of Biogeography* 29(12): 1595–1618.

Sundar Rajan, K. 2006. *Biocapital: The Constitution of Postgenomic Life*. Durham, NC: Duke University Press.

Swyngedouw, E. 1999. "Modernity and Hybridity: Nature, *Regeneracionismo*, and the Production of the Spanish Waterscape, 1890–1930." *Annals of the Association of American Geographers* 89(3): 443–65.

———. 2010. "Apocalypse Forever? Post-political Populism and the Spectre of Climate Change." *Theory, Culture & Society* 27(2–3): 213–32.

———. 2011. "The Trouble with Nature: Ecology as the New Opium for the Masses." In *The Ashgate Research Companion to Planning Theory*, edited by J. Hillier and P. Healey, 299–318. Ashgate: Aldershot.

Szabo, M. 2010. "Kyoto May Push Factories to Pollute More: UN Report." Reuters, July 2.

Szersynski, B. 2010. "Reading and Writing the Weather: Climate Technics and the Moment of Responsibility." *Theory, Culture & Society* 27(2–3): 9–30.

Taussig, M. 1987. *Shamanism, Colonialism and the Wild Man: A Study in Terror and Healing*. Chicago: University of Chicago Press.

Taylor, M. 2012. "Innovation under Cap and Trade Programs." *Proceedings of the National Academy of Sciences* 109(13): 4804–9.

TEEB. 2009. "The Economics of Ecosystems and Biodiversity for National and International Policy Makers—Summary: Responding to the Value of Nature." http://www.teebweb.org/LinkClick.aspx?fileticket=I4Y2nqqIiCg%3D.

———. 2010. "The Economics of Ecosystems and Biodiversity: Mainstreaming the Economics of Nature: A Synthesis of the Approach, Conclusions and Recommendations of TEEB." http://www.teebweb.org/TEEBSynthesisReport/tabid/29410/Default.aspx.

Terborgh, J. 1999. *Requiem for Nature*. Washington, DC: Island Press.

Tett, G. 2009. *Fool's Gold*. New York: Simon and Schuster.

Thomas, N. 1991. *Entangled Objects: Exchange, Material Culture, and Colonialism in the Pacific*. Cambridge, MA: Harvard University Press.

Thompson, E. P. 1990. *Customs in Common*. New York: Free Press.

Thrift, N. 2005. *Knowing Capitalism*. London: Sage.

Tsing, A. L. 2005. *Friction: An Ethnography of Global Connection*. Princeton, NJ: Princeton University Press.

Turner, G. 2004. *Understanding Celebrity*. London: Sage.

Turner, G., F. Bonner, and D. P. Marshall. 2000. *Fame Games: The Production of Celebrity in Australia*. Cambridge: Cambridge University Press.

Turnhout, E., C. Waterton, K. Neves, and M. Buizer. 2013. "Rethinking Biodiversity: From Goods and Services to 'Living With.'" *Conservation Letters* 6:154–61.

Tylor, E. (1871) 1913. *Primitive Culture*. 2 vols. London: John Murray.

UNCSD (United Nations Conference on Sustainable Development). 2012. "The Future We Want." http://www.uncsd2012.0rg/ri020/futurewewant.html.

UNDP (United Nations Development Programme). 2004. *Partnerships for Conservation: Lessons from the "COMPACT" Approach for Co-managing Protected Areas and Landscapes*. New York: UNDP.

UNEP (United Nations Environment Programme). 2011. *Towards a Green Economy: Pathways to Sustainable Development and Poverty Eradication.* Nairobi: UNEP. http://www.unep.org/greeneconomy.

UNEP, Risoe Centre. 2010. CDM Pipeline (spreadsheet). http://cdmpipeline.org/.

Unruh, G. C. 2000. "Understanding Carbon Lock-In." *Energy Policy* 28(12): 817–30.

UNWTO (United Nations World Tourism Organization). 2012. *Tourism Highlights 2011.* Madrid: UNWTO.

Valera, F. J. 1992. *Ethical Know-How: Action, Wisdom, and Cognition.* Stanford, CA: Stanford University Press.

Venturello, M. 1907. *Manners and Customs of the Tagbanuas and Other Tribes of the Island of Palawan, Philippines.* Smithsonian Miscellaneous Collections 48, pt. 3:514–58.

Virno, P. 1996. "The Ambivalence of the General Intellect." Translated by M. Turits. In *Radical Thought in Italy: A Potential Politics,* edited by M. Hardt and P. Virno, 265–73. Minneapolis: University of Minnesota Press.

Vitug, M. 1993. *The Politics of Logging: Power from the Forest.* Manila: Philippine Center for Investigative Journalism.

———. 2000. "Forest Policy and National Politics." In *Forest Policy and Politics in the Philippines: The Dynamic of Participatory Conservation,* edited by P. Utting, 11–40. Quezon City: Anteneo de Manila University Press.

Viveiros de Castro, E. 1998. "Cosmological Deixis and Amerindian Perspectivism." *Journal of the Royal Anthropological Institute* 4(3): 469–88.

———. 2004. "Exchanging Perspectives: The Transformation of Objects into Subjects in Amerindian Ontologies." *Common Knowledge* 10(3): 463–84.

Wacquant, L. 2012. "Three Steps to a Historical Anthropology of Actually Existing Neoliberalism." *Social Anthropology* 20(1): 66–79.

Walker, S., A. Brower, T. Stephens, and W. Lee. 2009. "Why Bartering Biodiversity Fails." *Conservation Letters* 2(4): 149–57.

Walsh, B. 2012. "Nature Is Over." *Time Magazine,* March 12, 83–85.

Wark, M. 1994. *Virtual Geography: Living with Global Media Events.* Bloomington: Indiana University Press.

Warner, K. 1979. "Walking on Two Feet: Tagbanuwa Adaptation to Philippine Society." Ph.D. diss., Anthropology Department, University of Hawaii.

Warren, C. 1977. *The Batak of Palawan: A Culture in Transition.* Chicago: University of Chicago Press.

WBCSD (World Business Council for Sustainable Development). 2011. "Guide to Corporate Ecosystem Valuation: A Framework for Improving Corporate Decision-Making." World Business Council for Sustainable Development. http://www.wbcsd .org/work-program/ecosystems/cev.aspx.

WCED (World Commission on Environment and Development). 1987. *Our Common Future: The Report of the World Commission on Environment and Development.* Oxford: Oxford University Press.

Weber, M. (1904) 2001. *The Protestant Ethic and the Spirit of Capitalism.* London: Routledge.

———. (1922) 1978. *Economy and Society: An Outline of Interpretive Sociology.* 2 vols. 2nd English ed., edited by Guenther Roth and Claus Wittich. Berkeley: University of California Press. [Translation of *Wirtschaft und Gesellschaft: Grundriss der verstehenden Soziologie*, based on 4th German ed., 1956; original German version published in 1922.]

Weberling, B. 2010. "Celebrity Charity: A Historical Case Study of Danny Thomas and St. Jude Children's Research Hospital, 1962–1991." *Prism* 7(2): 1–15.

Wells, M. P., and K. Brandon 1992. *People and Parks: Linking Protected Area Management with Local Communities*. Washington, DC: World Bank.

Wells, M. P., and T. O. McShane. 2004. "Integrating Protected Area Management with Local Needs and Aspirations." *Ambio* 33(8): 513–19.

West, P. 2006. *Conservation Is Our Government Now: The Politics of Ecology in Papua New Guinea*. Durham, NC: Duke University Press.

———. 2010. "Making the Market: Specialty Coffee, Generational Pitches, and Papua New Guinea." *Antipode* 42(3): 690–718.

West, P., and J. C. Carrier. 2004. "Ecotourism and Authenticity: Getting Away from It All?" *Current Anthropology* 45(4): 483–98.

Wheeler, W. 2010. "Gregory Bateson and Biosemiotics: Transcendence and Animism in the 21st Century." *Green Letters: Studies in Ecocriticism* 13(Winter): 35–54, special issue, "Ecophenomenology and Practices of the Sacred."

White, B. 1989. "Problems in the Empirical Analysis of Agrarian Differentiation." In *Agrarian Transformations*, edited by G. Hart, A. Turton, and B. White, 15–30. Berkeley: University of California Press.

White, B., S. Borras, Jr., R. Hall, I. Scoones, and W. Wolford. 2012. "The New Enclosures: Critical Perspectives on Corporate Land Deals." *Journal of Peasant Studies* 39(3/4): 619–47.

Willerslev, R. 2012. "Laughing at the Spirits in North Siberia: Is Animism Being Taken Too Seriously?" *e-flux* 36. http://www.e-flux.com/journal/laughing-at-the-spirits-in -north-siberia-is-animism-being-taken-too-seriously/ Accessed 6 January 2013.

Williams, C. 2005. *A Commodified World? Mapping the Limits of Capitalism*. London: Zed Books.

———. 2010. *Ecology and Socialism: Solutions to Capitalist Ecological Crisis*. Chicago: Haymarket Books.

Williams, R. (1976) 1983. *Keywords: A Vocabulary of Culture and Society*. Oxford: Oxford University Press.

Willis, R., and P. Curry. 2004. *Astrology, Science and Culture: Pulling Down the Moon*. Oxford: Berg.

Wilshusen, P. 2009. "Shades of Social Capital: Elite Persistence and the Everyday Politics of Community Forestry in Southeastern Mexico." *Environment and Planning A* 41:389–406.

———. 2010. "The Receiving End of Reform: Everyday Responses to Neoliberalization in Southeastern Mexico." *Antipode* 42(3): 767–99.

Wilson, A. 1993. *The Culture of Nature: North American Landscapes from Disney to the "Exxon Valdez."* Cambridge: Blackwell Publishers.

Wilson, J. 2012. "The *Jouissance* of Philanthrocapitalism." Paper presented at the symposium "Capitalism, Democracy, and Celebrity Advocacy," University of Manchester, UK, June 19–20.

Wolff, R. D. 2012. *Democracy at Work: A Cure for Capitalism*. Chicago: Haymarket Books.

World Bank. 1990. *World Development Report 1990*. New York: Oxford University Press.

————. 1997. *Expanding the Measure of Wealth: Indicators of Environmentally Sustainable Development*. Washington, DC: World Bank. http://info.worldbank.org/etools/docs/library/110128/measure.pdf.

Wunder, S. 2005. *Payments for Environmental Services: Some Nuts and Bolts*. CIFOR Occasional Paper 42. Jakarta: CIFOR.

————. 2008. "Payments for Environmental Services and the Poor: Concepts and Preliminary Evidence." *Environment and Development Economics* 13:279–97.

Young, A. 2011. "Putting the Sacred Cow Out to Pasture." Unpublished manuscript.

Yusoff, K. 2012. "Aesthetics of Loss: Biodiversity, Banal Violence and Biotic Subjects." *Transactions of the Institute of British Geographers* 37(4): 578–92.

Zbaracki, M. J. 2004. *Pricing Structure and Structuring Price*. Philadelphia: University of Pennsylvania.

Zelizer, V. 1995. *The Social Meaning of Money*. New York: Basic Books.

Žižek, S. 1989. *The Sublime Object of Ideology*. London: Verso.

————. 1997. *The Plague of Fantasies*. London: Verso.

————. 2008. *In Defense of Lost Causes*. London: Verso.

————. 2009. *First as Tragedy, Then as Farce*. London: Verso.

Contributors

Dan Brockington is a professor of conservation and development at the Institute for Development Policy and Management, University of Manchester, UK. His research covers diverse aspects of conservation, development, and celebrity. His recent books are *Celebrity and the Environment* (2009) and *Nature Unbound* (2008, with Rosaleen Duffy and Jim Igoe).

Bram Büscher is an associate professor of environment and sustainable development at the Institute of Social Studies, Erasmus University Rotterdam, the Netherlands, and visiting associate professor at the Department of Geography, Environmental Management and Energy Studies of the University of Johannesburg, South Africa. He is the author of *Transforming the Frontier: Peace Parks and the Politics of Neoliberal Conservation in Southern Africa* (2013).

Catherine Corson is an assistant professor of environmental studies at Mount Holyoke College. Her research focuses on neoliberal conservation, foreign aid politics, international environmental governance, institutional ethnography, and political ecology. She received her doctorate in 2008 from the University of California at Berkeley, and her dissertation analyzed how state and nonstate actors negotiate US environmental foreign aid to Madagascar across local, national, and international scales. Currently, she works with an international team to study, using collaborative event ethnography, the coproduction of conservation knowledge and relations of environmental governance. Prior to her academic career, she worked for ten years as an environment and development policy analyst and consultant.

Wolfram Dressler loves studying the particular and the odd in forest settings, things that are unique, deserving of attention, and deserving of survival. He has done so in settings as diverse as Laos, the Philippines, the Caribbean, South Africa, and the western Arctic. He is an ARC Future Fellow in the School of Land and Environment at the University of Melbourne, Australia, and the author of *Old Thoughts in New Ideas: State Conservation Measures, Livelihood and Development on Palawan Island.*

Robert Fletcher is an associate professor of natural resources and sustainable development at the United Nations–mandated University for Peace in Costa Rica. His research, conducted in North, Central, and South America, explores how culturally specific approaches to human-environment relations inform patterns of resource use and the contestation among these. He is the author of *Romancing the Wild: Cultural Dimensions of Ecotourism* (2014).

Jim Igoe is an associate professor of anthropology and a Mellon Fellow in the Institute of the Humanities and Global Culture at the University of Virginia. His research concerns relationships between indigenous peoples and national parks, the neoliberalization of nature conservation, and the anthropology of noncapitalist alternatives. He is the author of *Conservation and Globalization: A Study of Indigenous Peoples and National Parks from East Africa to South Dakota* and a proud member of the VIVA Collective.

Larry Lohmann has been an environmental activist in Thailand and Europe for over twenty-five years. He works with the Corner House, a small advocacy and research organization based in Dorset, UK.

Ken MacDonald teaches in the Department of Geography, the program in International Development Studies, and the Centre for Diaspora and Transnational Studies at the University of Toronto. His research interests focus on the cultural politics of development, the social construction of vulnerability, the organizational and institutional politics of biodiversity conservation policy and practice, and transnational geographies of consumption. Recent publications include essays in *Cultural Geographies*, *Geoforum*, and *Conservation and Society*. His current book project, *Geographies of Intervention: Postcolonialism and the Politics of Development in Northern Pakistan*, is based on twenty years of fieldwork in Pakistan's Northern Areas.

Frank Matose is a senior lecturer in the Department of Sociology at the University of Cape Town. His interests are in natural resource commons with a particular focus on southern Africa, placing emphasis on the intersection of local people, the state, capital, forest conservation, and protected areas. Interests in these areas are informed by intellectual projects around environmental governance, social justice, knowledge, and power. He recently coedited the book *Coping amidst Chaos: Studies on Adaptive Collaborative Management from Zimbabwe* with Alois Mandondo and Ravi Prabhu.

Sian Sullivan is professor of environment and culture at the School of Society, Enterprise and Environment at Bath Spa University, UK. She has conducted long-term research on changing people–landscape relationships in northwest Namibia, as well as on the politics of subjectivity in the global justice movement. She is currently studying financialization processes in biodiversity conservation. She is coeditor of *Political Ecology: Science, Myth and Power* (with Edward Arnold) and has published in a range of journals and edited collections, with recent articles appearing in *Antipode, New Formations, Environmental Conservation, Globalizations, Capitalism Nature Socialism, Radical Anthropology, Forum for Development Studies,* and *Conservation and Society*. She is working on a book entitled *Creating Earth Incorporated? Nature, Finance, Values.*

Peter R. Wilshusen is an associate professor of environmental studies and an affiliate faculty member in the Department of Geography at Bucknell University in Lewisburg, Pennsylvania. He currently serves as the executive director of the Bucknell University Environmental Center and holds the David and Patricia Ekedahl Professorship in Environmental Studies.

Index